WATER RESOURCE PLANNING, DEVELOPMENT AND MANAGEMENT

PHARMACEUTICALS IN THE U.S. WATER SUPPLY

Water Resource Planning, Development and Management

Additional books in this series can be found on Nova's website under the Series tab.

Additional E-books in this series can be found on Nova's website under the E-book tab.

Environmental Remediation Technologies, Regulations and Safety

Additional books in this series can be found on Nova's website under the Series tab.

Additional E-books in this series can be found on Nova's website under the E-book tab.

WATER RESOURCE PLANNING, DEVELOPMENT AND MANAGEMENT

PHARMACEUTICALS IN THE U.S. WATER SUPPLY

ED T. BLEVINS
AND
SHERMAN E. RUSH
EDITORS

Nova Science Publishers, Inc.
New York

For permission to use material from this book please contact us:
Telephone 631-231-7269; Fax 631-231-8175
Web Site: http://www.novapublishers.com

Library of Congress Cataloging-in-Publication Data

ISBN 978-1-61942-408-1

Published by Nova Science Publishers, Inc. † New York

CONTENTS

PREFACE

In response to increasing information arising from the scientific community, the news media reported that pharmaceuticals have been detected in the drinking water of 24 major metropolitan areas across the United States. These reports raised concerns about the potential impact of these pharmaceuticals on human health and the environment. Pharmaceuticals are a particular concern because they are designed to interact with human or animal physiology. Much is known about the therapeutic uses of pharmaceuticals, but little is known about their potential risk to human health from long-term exposure through drinking water. This book examines the research reported in the detection of pharmaceuticals in the U.S. water supply.

Chapter 1- Drinking water in some metropolitan areas contains concentrations of pharmaceuticals, raising concerns about their potential impact on human health. The Safe Drinking Water Act (SDWA) authorizes the Environmental Protection Agency (EPA) to regulate contaminants, including pharmaceuticals, in public drinking water systems if they may adversely affect human health among other criteria. Pharmaceuticals may enter drinking water supplies from several pathways, including discharge from wastewater facilities. GAO was asked to provide information on the (1) extent to which pharmaceuticals occur in drinking water and their effects, if any, on human health; (2) U.S. and other countries' approaches to reducing their occurrence; and (3) challenges, if any, that EPA faces in determining whether to regulate pharmaceuticals. GAO reviewed federal and peer-reviewed reports, and surveyed a nonprobability sample of five U.S. programs designed to properly dispose of pharmaceuticals. We selected these programs based on geographic diversity and program characteristics. We also researched such programs in two countries, and interviewed scientists and agency officials.

Chapter 2- We are here today to conduct much-needed oversight on the presence of pharmaceuticals in our nation's water supplies.

Clean, safe drinking water is essential to all of us. It is especially important for children.

The National Academy of Sciences has found that children drink more water and eat more food in proportion to their body weight compared to adults. In addition, children's rapidly developing bodies, including their hormone systems that control their development, are especially vulnerable. Pregnant women also undergo a host of similarly delicate changes.

There are particular "windows of vulnerability" during our development when pregnant women and children may be especially susceptible to very low doses of some toxins.

Chapter 3- The observed presence of pharmaceuticals in the environment has prompted public interest regarding potential adverse ecological effects and potential contamination of drinking water. The interest has already increased public awareness of the ways we handle and dispose of our medications and has resulted in interest by industries in waste treatment technologies and best management practices that are most effective at removing pharmaceuticals and other trace organic chemicals from surface and ground waters and solid and liquid wastes.

The USGS studies a wide range of chemicals of emerging environmental concern. Human and veterinary pharmaceuticals are only one class of chemicals in a new group of contaminants that enter the environment via human and animal wastes. These contaminants include many chemicals used in our homes, businesses, and industries, including detergents, fragrances, fire retardants, disinfectants, plastics, and insect repellants.

Chapter 4- I appreciate the opportunity to describe EPA's actions to evaluate the potential risks to human health and aquatic life posed by trace amounts of pharmaceuticals in water, and to identify measures to minimize their occurrence in water. The Agency is committed to undertaking a scientific approach in evaluating the risks associated with contaminants in our environment so that we can take the necessary action to ensure clean and safe water.

EPA is concerned about the detection of a number of pharmaceuticals and personal care products in our water. EPA has been actively working with federal agencies and state and local government partners to better understand the implications of emerging contaminants such as pharmaceuticals, endocrine disrupting chemicals, and personal care products detected in drinking water, wastewater, surface water and ground water. We continue to evaluate their routes of exposure, levels of exposure, and potential effects on public health and aquatic life.

Chapter 5- Compounds such as nicotine, caffeine, and aspirin that are designed to influence our body's normal chemistry have been identified as environmental contaminants since the 1980s, moving from sewage and human waste into waterways. As our use of pharmaceuticals increases, it is logical to expect them to turn up in our environment. Although the levels reported to contaminate our waterways are much lower than therapeutic doses, it would be naïve to think of this as 'safe', knowing that the agents are chemically reactive in our bodies, and that we are exposed daily over a life-time to multiple compounds in unknown combinations.

When a medical professional prescribes a drug, they are considering the patient's health status, age, gender, nutritional status, and any other drugs that may cross-react. For example, a woman who was at risk for breast cancer would not want to be exposed to high levels of estrogenic compounds. A pregnant woman would never knowingly expose her fetus to chemicals that cause birth defects such as antiepileptic drugs. A doctor would never knowingly prescribe toxic chemotherapy agents to a healthy person. Yet, all these things and more are in our Nation's drinking water. The Associated Press reported that pharmaceutical residues were detected in the drinking water of 24 major metropolitan areas across the country serving 41 million people. Detected drugs included antibiotics, anti-convulsants, and mood stabilizer drugs. These results were supported by findings of the U.S. Geological Survey that found organic wastewater contaminants and pharmaceuticals in 80% of sampled streams- including antibiotics, hypertensive and cholesterol-lowering drugs, antidepressants, analgesics, steroids, caffeine, and reproductive hormones.

Chapter 6- I am the Deputy Vice President for Regulatory Affairs at the Pharmaceutical Research and Manufacturers of America (PhRMA), a trade association representing the leading research-based pharmaceutical and biotechnology companies. PhRMA member companies invested an estimated $44.5 billion in 2007 for innovative biomedical research to discover and develop new medicines that meet medical needs.

It is not a new development that drinking water may contain trace levels of pharmaceuticals as well as other household chemicals. What is new is that we now have more sensitive analytical methods to detect these substances at very low levels.

Reports of the presence of trace levels of pharmaceuticals in the water sparked the creation of a Pharmaceuticals in the Environment (PIE) Task

Force within PhRMA in the late 1990s. The PIE Task Force is familiar with essentially all of the reports published worldwide of pharmaceuticals in water sources, including testing conducted by the U.S. Geological Survey (USGS), routine municipal water testing, testing by scientists from Europe to Asia, and even high school science students whose projects were on display in school science fairs.

In Pharmaceuticals in the U.S. Water Supply ISBN: 978-1-61942 -408-1
Editors: E. T. Blevins and S. E. Rush

Chapter 1

ENVIRONMENTAL HEALTH: ACTION NEEDED TO SUSTAIN AGENCIES' COLLABORATION ON PHARMACEUTICALS IN DRINKING WATER*

United States Government Accountability Office

WHY GAO DID THIS STUDY

Drinking water in some metropolitan areas contains concentrations of pharmaceuticals, raising concerns about their potential impact on human health. The Safe Drinking Water Act (SDWA) authorizes the Environmental Protection Agency (EPA) to regulate contaminants, including pharmaceuticals, in public drinking water systems if they may adversely affect human health among other criteria. Pharmaceuticals may enter drinking water supplies from several pathways, including discharge from wastewater facilities. GAO was asked to provide information on the (1) extent to which pharmaceuticals occur in drinking water and their effects, if any, on human health; (2) U.S. and other countries' approaches to reducing their occurrence; and (3) challenges, if any, that EPA faces in determining whether to regulate pharmaceuticals. GAO

* This is an edited, reformatted and augmented version of The United States Government Accountability Office publication, Report to Congressional Requesters GAO-11-346, dated August 2011.

reviewed federal and peer-reviewed reports, and surveyed a nonprobability sample of five U.S. programs designed to properly dispose of pharmaceuticals. We selected these programs based on geographic diversity and program characteristics. We also researched such programs in two countries, and interviewed scientists and agency officials.

What GAO Recommends

GAO recommends that the Administrator of EPA establish a workgroup or other formal mechanism to coordinate research on pharmaceuticals and other contaminants in drinking water. EPA agreed with the recommendation.

What GAO Found

Research has detected pharmaceuticals in the nation's drinking water. National and regional studies by the U.S. Geological Survey, EPA, and others have detected pharmaceuticals in source water, treated drinking water, and treated wastewater; but the full extent of occurrence is unknown. The concentrations detected for any one pharmaceutical were measured most frequently in parts per trillion. Research has not determined the human health effects of exposure to these concentrations of pharmaceuticals in drinking water. However, federal research has demonstrated the potential impact to human health from exposure to some pharmaceuticals found in drinking water, such as antibiotics and those that interfere with the functioning and development of hormones in humans.

Some states and local governments as well as the Drug Enforcement Administration have taken actions that could reduce the extent to which pharmaceuticals occur in drinking water. These efforts have primarily been through drug take-back programs to encourage proper control and disposal of pharmaceuticals. Additional efforts have been adopted in Europe following the European Union's directive in 2004 requiring member states to have appropriate collection systems for unused or expired medicinal products. In addition to collection systems, Sweden also encourages actions such as writing small initial prescriptions to reduce the amount of pharmaceuticals that are disposed of if patients switch to a different pharmaceutical course.

EPA faces challenges in obtaining sufficient occurrence and health effects data on pharmaceuticals and other contaminants in drinking water to support analyses and decisions to identify which, if any, pharmaceuticals should be regulated under SDWA. EPA is collaborating with the Food and Drug Administration and U.S. Geological Survey on research to help obtain such data but these efforts are largely informal. EPA officials said there is no formal mechanism, such as a long-term strategy or formal agreement, to manage and sustain these collaborative efforts. A recently expired interagency workgroup, which EPA co-chaired, initiated work on a research strategy to identify opportunities that will enhance collaborative federal efforts on pharmaceuticals in the environment, but its draft report did not contain key details about how the agencies will coordinate such collaborative efforts. GAO previously identified key practices for enhancing and sustaining collaboration among federal agencies, some of which may help clarify such coordination, such as establishing the roles and responsibilities of collaborating agencies; leveraging their resources; and establishing a process for monitoring, evaluating, and reporting to the public the results of the collaborative research efforts.

ABBREVIATIONS

ADI	acceptable daily intake
BPA	bisphenol A
CENRS	Committee on Environment, Natural Resources, and Sustainability
DEA	Drug Enforcement Administration
DOI	Department of the Interior
DOJ	Department of Justice
EDC	endocrine disrupting chemical
EPA	Environmental Protection Agency
EU	European Union
FDA	Food and Drug Administration
HHS	Department of Health and Human Services
OSTP	Office of Science and Technology Policy
NSTC	National Science and Technology Council
PiE	Pharmaceuticals in the Environment
RCRA	Resource Conservation and Recovery Act of 1976
RUM	Return Unwanted Medicines

SDWA Safe Drinking Water Act
USGS U.S. Geological Survey

The Honorable Brad Miller
Ranking Member
Subcommittee on Energy and Environment
Committee on Science, Space and Technology
House of Representatives

The Honorable Edward J. Markey
House of Representatives

In 2008, in response to increasing information arising from the scientific community, the news media reported that pharmaceuticals had been detected in the drinking water of 24 major metropolitan areas across the United States.[1] The concentrations detected were measured most frequently in parts per trillion. The reports raised concerns about the potential impact of these pharmaceuticals on human health and the environment. Pharmaceuticals are a particular concern because they are designed to interact with human or animal physiology.[2] Much is known about the therapeutic uses of pharmaceuticals, but little is known about their potential risk to human health from long-term exposure through drinking water. According to scientists, pharmaceuticals may enter the environment and ultimately drinking water supplies in various ways, such as through the elimination of human and animal waste, disposal of unused medicines down the toilet or drain, veterinary drug usage, hospital waste disposal, and industrial discharges.

Federal, state, and local governments, and other countries have taken actions to reduce the extent to which pharmaceuticals enter the environment through programs that encourage the proper disposal of unused and expired pharmaceuticals. These programs are known as take-back programs and in the United States they are also tied to efforts to reduce drug abuse or accidental poisoning by removing expired medicines from homes.

The Environmental Protection Agency (EPA) has the authority to regulate some pharmaceuticals under several statutes, including its responsibility for regulating the nation's drinking water under the Safe Drinking Water Act (SDWA). Pharmaceuticals are regulated by other agencies, including the Food and Drug Administration (FDA) and in some cases the Drug Enforcement Administration (DEA), but not as contaminants in drinking water.

To help understand the human health and ecological effects of pharmaceuticals in the environment, a workgroup of federal scientists was established in 2006 to identify and prioritize research to better understand the risk from pharmaceuticals in the environment and to recommend areas for federal collaboration to address those priorities. The workgroup, known as the Pharmaceuticals in the Environment (PiE) workgroup, produced a draft report in the spring of 2009 that was never finalized or publicly released.

In this context, you asked us to review the scientific literature and assess efforts to address pharmaceuticals in drinking water. Our objectives were to (1) provide information on the extent to which pharmaceuticals occur in drinking water and the effects, if any, that their occurrence has on human health; (2) describe the approaches taken in the United States and in other countries to reduce the extent to which pharmaceuticals occur in drinking water; and (3) identify challenges, if any, that EPA faces in determining whether any pharmaceuticals should be regulated under SDWA, actions EPA is taking to address these challenges, and options for addressing such challenges in the future.

To identify the extent to which pharmaceuticals occur in drinking water and their potential effects on human health, we reviewed federal and peer-reviewed reports, including (1) studies by the U.S. Geological Survey (USGS), (2) articles in scientific journals, and (3) the PiE workgroup's draft report. We also interviewed scientists and other officials from federal agencies, as well as representatives from the pharmaceutical industry. Additionally, we attended an October 2009 academic conference on hormones and related compounds in the environment. To describe the approaches taken in the United States and in other countries to reduce the extent to which pharmaceuticals occur in drinking water; we reviewed literature and spoke with experts; from these efforts, we identified consumer take-back programs as the primary approach to reducing occurrence in the United States. We selected a nonprobability sample of five U.S take-back programs to provide geographic and program diversity; the information from these programs is not generalizeable to all take-back programs and offers examples of how these programs can operate . We also chose to describe efforts by Sweden and Australia to reduce the occurrence of pharmaceuticals in drinking water. We selected Sweden because it engages in a variety of activities to reduce the occurrence of pharmaceuticals in drinking water. We selected Australia because it has a national take-back program. To identify challenges, if any, that EPA faces in determining whether any pharmaceuticals should be regulated under SDWA, actions EPA is taking to address these challenges, and options for addressing such challenges in the future, we reviewed documentation related to EPA's implementation of SDWA and interviewed EPA and other federal officials knowledgeable about challenges EPA is facing in implementing SDWA. We also reviewed the PiE workgroup's 2009 draft report and interviewed workgroup members and Office of Science and Technology Policy (OSTP) officials. We also reviewed our own work on practices that can help enhance and sustain interagency collaboration. A more detailed description of our scope and methodology is presented in appendix I.

We conducted this performance audit from January 2010 through August 2011 in accordance with generally accepted government auditing standards. Those standards require that we plan and perform the audit to obtain sufficient, appropriate evidence to provide a reasonable basis for our findings and conclusions based on our audit objectives. We believe that the evidence obtained provides a reasonable basis for our findings and conclusions based on our audit objectives.

Background

This section presents information on (1) the ways in which pharmaceuticals may enter drinking water, (2) pharmaceuticals in drinking water as a contaminant of emerging concern, (3) the degree to which relevant environmental statutes regulate pharmaceuticals, and (4) the establishment of the PiE workgroup.

Pharmaceuticals May Enter Drinking Water through Different Pathways

Scientists have identified numerous pathways by which pharmaceuticals may enter the environment and ultimately drinking water supplies. According to USGS scientists, the main source of human pharmaceuticals in the environment is likely treated wastewater from households, industry, and commercial facilities. Biosolids from wastewater treatment plants applied to land as fertilizer may also be a source of human pharmaceuticals in the environment. Septic systems may be a source of human pharmaceuticals in ground water. A potential source of veterinary pharmaceuticals is agricultural facilities where large numbers of food-producing animals (such as chickens, cattle, and swine) are treated with pharmaceuticals. The pharmaceuticals enter the environment either directly from waste storage structures as a result of accidents or weather conditions, or through the application of manure and liquid waste to croplands. Figure 1 illustrates the different pathways by which pharmaceuticals may enter drinking water supplies.

Source: GAO.

Figure 1. Examples of Pathways through which Pharmaceuticals May Enter Drinking Water Supplies.

Pharmaceuticals in Drinking Water Are a Contaminant of Emerging Concern

EPA considers pharmaceuticals in drinking water to be a contaminant of emerging concern (also called emerging contaminants). The term is not defined in regulation, and EPA does not maintain a list of contaminants that are considered contaminants of emerging concern. In this report, the term refers to a wide range of contaminants for which the risk to human health and the environment associated with their presence, frequency of occurrence, or source may not be known. In some cases, the release of contaminants of emerging concern into the environment has likely occurred for a long time but may not have been recognized until new detection methods were developed.

In other cases, the synthesis of new chemicals or changes in the use and disposal of existing chemicals can create new sources of contaminants of emerging concern. Other contaminants of emerging concern can include personal care products (e.g., sunscreen, antibacterial soap, synthetic musks); chemicals used in industry (e.g., flame retardants, stain resistant coatings); and chemicals used in agriculture (e.g., pesticides that may act as endocrine disrupting chemicals (EDC)).[3]

The Degree to which Relevant Environmental Statutes Regulate Pharmaceuticals

Most pharmaceuticals are not currently regulated under EPA programs implementing key environmental laws. SDWA, the Resource Conservation and Recovery Act of 1976 (RCRA) and the Clean Water Act provide EPA with authority to regulate pharmaceuticals meeting certain criteria in drinking water, waste, and wastewater discharges.

The Safe Drinking Water Act

Under SDWA, EPA is authorized to regulate contaminants, including pharmaceuticals, meeting certain criteria in public drinking water systems. In 1996, Congress amended SDWA to require EPA to select for consideration those unregulated contaminants that present the greatest public health concern, evaluate their occurrence and the potential health risks associated with them, and decide whether a regulation is needed for at least five contaminants every 5 years. This regulatory determination process includes EPA's publication in the *Federal Register* of a preliminary decision on whether the agency will propose a drinking water regulation for each contaminant evaluated—called a preliminary regulatory determination—and provides for a public comment period, followed by a final decision, or regulatory determination, also published in the *Federal Register*. The 1996 amendments also require EPA to identify and publish a list every 5 years of unregulated contaminants for drinking water that may require regulation—called the Contaminant Candidate List. The Administrator must decide whether to regulate at least five of the contaminants on the candidate list every 5 years.[4] These decisions are called regulatory determinations. SDWA specifies that EPA is to regulate a contaminant if the Administrator determines that:

- the contaminant may have an adverse effect on the health of persons;
- the contaminant is known to occur or there is a substantial likelihood that the contaminant will occur in public water systems with a frequency and at levels of public health concern; and
- in the sole judgment of the Administrator, regulation of such contaminant presents a meaningful opportunity for health risk reduction for persons served by public water systems.

Since 1996, EPA has completed two regulatory determination cycles—in 2003 and 2008. During this time, EPA conducted 20 regulatory determinations and found that none met the criteria requiring regulation. In 2011, EPA made an out-of-cycle regulatory determination, concluding that perchlorate, an ingredient in rocket fuel and other products that can interfere with the normal functioning of the thyroid gland, met the criteria requiring regulation.[5] EPA has made no regulatory determinations for pharmaceuticals. EPA published the third candidate list in October 2009 but has not yet made any regulatory determinations or completed the third regulatory determination cycle.[6]

To determine which contaminants to include on the third candidate list, EPA developed a multistep process, based on available data, to characterize occurrence and adverse health risks a contaminant may pose to consumers of public water systems.[7] Starting with a list of almost 26,000 unique chemicals, EPA identified a universe of about 6,000 potential drinking water contaminants for consideration based on the availability of occurrence and health effects data. Of these, 287 were pharmaceuticals. Then, using the available data, EPA employed successively more detailed evaluations—as well as expert opinions and comments from the public—to identify the 116 contaminants that it included on the third candidate list—12 of these contaminants are pharmaceuticals.8 Table 1 identifies the 12 pharmaceuticals.

In a May 2011 report, we identified systemic limitations in EPA's implementation of the 1996 amendments' requirements for determining whether additional contaminants in public drinking water warrant regulation and made 17 recommendations to EPA for implementing the requirements in a way that better assures the public of safe drinking water. Among other things, we recommended that EPA (1) develop criteria and a process for identifying those contaminants on its candidate list that present the greatest public health concern and (2) develop a coordinated process for obtaining both the occurrence and health effects data that may be needed for the agency to make informed regulatory determinations on these priority contaminants.[9]

Table 1. Pharmaceuticals on the Third Contaminant Candidate List

Pharmaceutical	Use
17alpha-estradiol	An estrogenic hormone used in pharmaceuticals
Equilenin	An estrogenic hormone used in pharmaceuticals
Equilin	An estrogenic hormone used in pharmaceuticals
Erythromycin	Used in pharmaceutical formulations as an antibiotic
Estradiol (17-beta estradiol)	An estrogenic hormone used in pharmaceuticals
Estriol	An estrogenic hormone used in veterinary pharmaceuticals
Estrone	An estrogenic hormone used in veterinary and human pharmaceuticals
Ethinyl estradiol (17-alpha ethynylestradiol)	An estrogenic hormone used in veterinary and human pharmaceuticals
Mestranol	An estrogenic hormone used in veterinary and human pharmaceuticals
Nitroglycerin	Used in pharmaceuticals, in the production of explosives, and in rocket propellants
Norethindrone (19-norethisterone)	A progresteronic hormone used in pharmaceuticals
Quinoline	Used in the production of other substances, as a pharmaceutical (antimalarial), andas a flavoring agent

Source: EPA.

Notes: According to FDA, estriol, estrone, ethinyl estradiol, and mestranol are not listed as approved for veterinary use in the United States but it is possible that these pharmaceuticals are or have been used as such, for example, for research purposes or as investigational new drugs. Although not identified as a use by EPA, according to FDA, estradiol and erythromycin are approved for veterinary use.

EPA did not agree to adopt these recommendations and generally took the position that no further steps are needed.

Resource Conservation and Recovery Act

RCRA established federal requirements and EPA regulatory authority for "cradle-to-grave" management of hazardous wastes, as well as a program for state oversight of nonhazardous solid waste with federal minimum regulations for landfills. RCRA and its implementing regulations establish several means by which waste may be deemed hazardous, including specifically being listed by EPA as a hazardous waste or by exhibiting one of the following four characteristics: toxicity, ignitability, corrosivity, or reactivity. According to EPA's August 2010 draft guidance and a proposed rule concerning management of hazardous pharmaceutical wastes in the *Federal Register*, more than 30 active pharmaceutical ingredients are considered listed hazardous wastes under RCRA.[10] In addition, other pharmaceuticals may be considered to be hazardous waste when disposed if they have certain characteristics (e.g., they are likely to leach concentrations of any 1 of 40 different toxic chemicals in amounts above the specified regulatory levels). Examples of these chemicals that have pharmaceutical uses include: arsenic, barium, cadmium, and chloroform. EPA has estimated that about 5 percent of all pharmaceutical waste is hazardous waste.[11]

The disposal of pharmaceuticals meeting the RCRA hazardous definition is generally subject to RCRA requirements, such as reporting, using a manifest, and disposing of the waste in approved ways, such as through hazardous waste incineration; however, household trash is exempted. Noting that implementing existing regulations may be difficult for healthcare facilities such as hospitals and nursing homes and that the streamlined requirements would help avoid mismanagement, in 2008 EPA proposed to add hazardous waste pharmaceuticals to the Universal Waste Rule, which simplifies RCRA requirements for certain hazardous wastes. Under the proposed rule, manifests would not be required and other requirements may be simplified. EPA estimated the rule could affect over 600,000 entities. According to EPA's Web site on the proposed rule, stakeholders commenting on the proposal expressed concerns that including hazardous pharmaceutical wastes under the Universal Waste Rule would eliminate some requirements, such as notification and use of a manifest, that currently apply to such wastes. EPA officials also told us the agency has begun considering additional regulatory options to address these and other issues but that EPA has no projected date for issuing a final rule.[12]

Clean Water Act

The Clean Water Act is the primary federal law concerning pollution of the nation's waters. Under the act, EPA is required to establish and revise national water quality criteria that accurately reflect the latest scientific knowledge about the effects of pollutants on aquatic life and human health. These criteria represent maximum concentrations that would not cause an unacceptable effect on aquatic life and represent the levels at which specific chemicals are not likely to adversely affect human health. Criteria are elements of state water quality standards, expressed as constituent concentrations, levels, or narrative statements, representing a quality of water that supports a particular use. When criteria are met, water quality will generally protect the designated use.[13] States, or in some instances EPA, use these criteria to adopt and revise water quality standards for designated uses—such as drinking, swimming, or fishing— for water bodies. States may use EPA's national criteria, modify them to site-specific criteria, or adopt other scientifically defensible criteria. States are required, as part of 3-year reviews, to adopt water quality standards for each of the toxic pollutants for which EPA has promulgated water quality criteria. Water quality standards play a critical role in the act's framework, potentially affecting effluent[14] limitations dictated by permits and requirements for state reporting and pollution control planning.

Regarding permits, EPA and delegated states administer the Clean Water Act's National Pollutant Discharge Elimination System program, which limits the types and amounts of pollutants that industrial and municipal wastewater treatment facilities may discharge into the nation's surface waters. Facilities such as municipal wastewater treatment plants and pharmaceutical plants require a permit if they discharge into surface waters. Certain agricultural facilities—known as concentrated animal feeding operations—also need a permit, but other agricultural operations do not. EPA and delegated states issue discharge permits that are to set conditions in accordance with technology-based effluent limitations EPA established for various categories of discharges. EPA has issued effluent limitation regulations for pharmaceutical manufacturing facilities as well as pretreatment regulations applicable when these facilities discharge into a publicly owned wastewater treatment plant. These regulations currently do not include limitations for any pharmaceutical constituents in wastewater; rather, the regulations set limitations for conventional pollutants, priority toxic pollutants, and selected nonconventional pollutants—mainly solvents used in manufacturing. Similarly, EPA's regulation for concentrated animal feeding operations does not contain specific limitations for veterinary pharmaceuticals.

At present, EPA has not developed specific water quality criteria under the Clean Water Act for most pharmaceuticals; hence, there are no water quality standards for most pharmaceuticals, and permits do not contain any limitations for them. EPA's current national criteria include one pollutant identified as being used as a pharmaceutical—lindane.[15,16] In January 2010, the Center for Biological Diversity, a nonprofit environmental organization, petitioned EPA to revise its water quality criteria for lindane, and to establish water quality criteria for 34 other pharmaceutical and personal care products. EPA told us the agency is considering the petition and expects to issue a response by mid 2011. If EPA were to establish water quality criteria for one or more additional pharmaceuticals, then states would need to adopt water quality standards reflecting the new or revised criteria, and the standards would be considered in permit decisions as well as in states' water quality management plans.

In August 2010, EPA's Office of Water released a draft guidance document for health care facilities, *Best Management Practices for Unused Pharmaceuticals at Health Care Facilities*. The nonbinding document recommends management practices, such as methods to reduce the quantity of unused pharmaceuticals, and explains applicable disposal requirements for those pharmaceuticals that are hazardous. EPA's goal for the guidance document is to keep pharmaceuticals out of U.S. waters, particularly by minimizing their disposal into sewers. According to agency officials, EPA expects to issue a final guidance document by the end of 2011.

Workgroup Addressing Pharmaceuticals in the Environment Was Established within the National Science and Technology Council

The PiE workgroup was established in 2006 by the Committee on Environment, Natural Resources, and Sustainability (CENRS), Toxics and Risk Subcommittee, an executive branch entity under the National Science and Technology Council (NSTC).[17] NSTC is a council of cabinet-level officials chaired by the President and managed by the Director of OSTP.[18] The purpose of the workgroup was to identify and prioritize research needed to better understand the risk from pharmaceuticals in the environment and to recommend areas for federal collaboration to address those priorities.[19] The workgroup, which was intended to be temporary, was staffed by scientists from eight federal agencies.[20] EPA, FDA, and USGS scientists served as co-chairs. In May 2009, the PiE workgroup produced a draft report but it was never finalized because of a disagreement between OSTP and the workgroup over what should be included in the final report.

Pharmaceuticals Have Been Found in Drinking Water, but Their Prevalence and Effects on Human Health Are Largely Unknown

Although research has confirmed the presence of pharmaceuticals in drinking water throughout the nation, the full extent of their occurrence is unknown. Research on the human health effects of exposure to these pharmaceuticals is largely unknown but the effects of some compounds have raised concern among some scientists, the public, and policy makers.

Research Has Confirmed the Presence of Some Pharmaceuticals in Drinking Water

Research has detected pharmaceuticals in the nation's drinking water. National and regional studies have generally detected pharmaceuticals in source water, treated drinking water, and treated wastewater; but the full extent of occurrence is unknown. The concentrations detected were measured most frequently in parts per trillion.

Studies Have Detected the Occurrence of Pharmaceuticals in Source Water

As part of its Toxic Substances Hydrology Program, USGS conducted four reconnaissance studies that were national in scope (national reconnaissance studies) to study the occurrence and distribution of emerging contaminants, including pharmaceuticals, in the environment. For each study, USGS chose to sample water from locations that it believed were more likely to have pharmaceuticals and other contaminants present.

One study specifically focused on untreated source water used by public drinking water systems. For example, samples were collected from wells and near surface water intakes that supplied the water systems. For this study, USGS collected water samples from 74 locations in 25 states and Puerto Rico in 2001. These locations provide drinking water to populations ranging from one family to over 8 million.[21] The study reported testing for the presence of 100 contaminants, including 36 pharmaceuticals.[22] USGS found that 53 of the 74 locations had one or more pharmaceuticals in the water, and 40 percent of the pharmaceuticals analyzed were detected at one or more of these locations. Figure 2 shows the location of sample sites and the sites at which USGS detected pharmaceuticals.

Figure 3 shows the pharmaceuticals that USGS reported detecting in its study of untreated sources of public drinking water.

The other three national reconnaissance studies that USGS conducted focused on (1) surface water,[23] (2) ground water,[24] and (3) stream sedimentation.[25] The four USGS national reconnaissance studies tested for a similar, but not identical, suite of pharmaceuticals and other contaminants and all of the studies reported detecting pharmaceuticals and other contaminants.

Source: GAO; map (MapArt).

Figure 2. Locations at which USGS Sampled Untreated Sources of Public Drinking Water and Those at which It Detected Pharmaceuticals.

In addition to its national studies, USGS has undertaken a number of local and regional studies as part of its reconnaissance effort to provide information on the sources, occurrence, and transport[26] of contaminants of emerging concern, including pharmaceuticals. These studies have reported similar results—finding pharmaceuticals in source water. For example, in a 2009 study, USGS, in cooperation with the Oregon Department of Environmental Quality and Deschutes County Environmental Health Division, collected and analyzed water samples from ground water near La Pine, Oregon.[27] The study reported detecting 8 of the 18 pharmaceuticals for which it tested. The study also reported testing for and finding other contaminants.

In addition to USGS, other research groups have conducted studies to detect pharmaceuticals and other contaminants in source water, with results that are similar to those of USGS. Specifically:

- The New York City Department of Environmental Protection reported finding pharmaceuticals and personal care products in the low, part-per-trillion range in a 2010 study of the Catskill, Croton, and Delaware untreated source waters that contributed to New York City's water supply.[28]

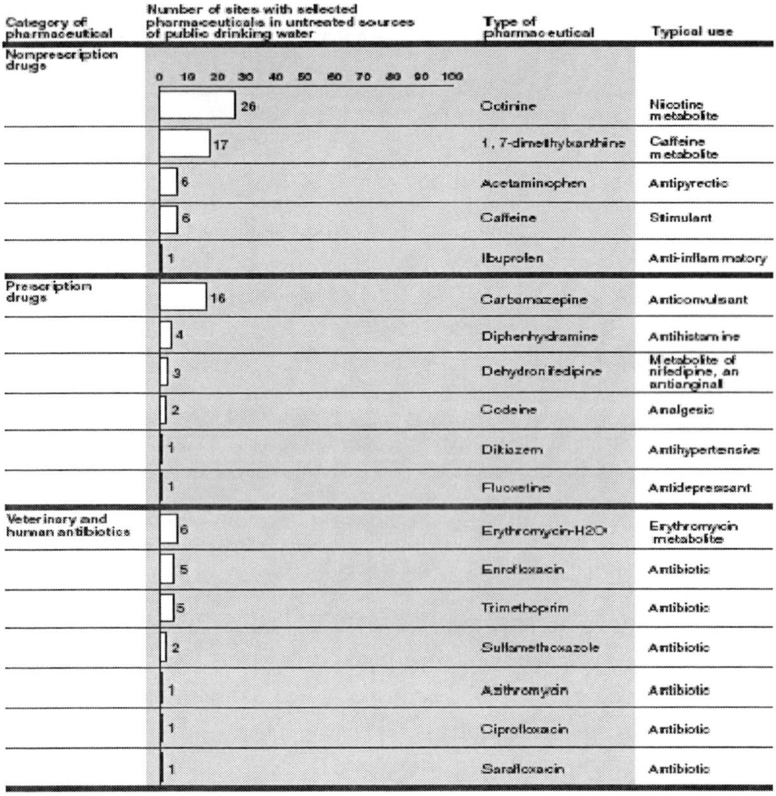

Source: GAO; map (MapArt).

Notes: A metabolite is a substance that is the product of biological changes to a chemical. For example, cotinine is the metabolite produced in the body after nicotine has been consumed. According to FDA, while caffeine, caffeine metabolites, and nicotine are in nonprescription drugs, caffeine and its metabolites are most likely from the urine of persons imbibing caffeine-containing beverages and nicotine metabolites are most likely from the urine of smokers. Also, antipyretic medication reduces fever.

Figure 3. Pharmaceuticals USGS Detected at Untreated Sources of Public Drinking Water.

- The National Water Research Institute funded a study testing for 50 contaminants such as pharmaceuticals and organic wastewater contaminants in three watersheds supplying drinking water to more than 25 million people in California.[29,30] The study analyzed 126 samples taken from 32 locations at various points in the watershed, including upstream and downstream from wastewater treatment plant discharges over a 1-year period, from April 2008 through April 2009. Overall, at least 1 contaminant was found in all but one of the samples. The study further reported that concentrations of contaminants were higher downstream of the wastewater treatment plants and concluded that the plant discharges were likely the main source of these contaminants in the environment.

Studies Have Also Detected Occurrence of Pharmaceuticals in Treated Drinking Water

Although USGS studies have focused on source water, other studies have detected pharmaceuticals and other emerging contaminants in treated drinking water. For example:

A 2008 study funded by the American Water Works Association Research Foundation and the WateReuse Foundation tested for 51 potential contaminants including 20 pharmaceuticals and pharmaceutical metabolites in drinking water in 19 drinking water treatment plants across the United States.[31,32] The study reported detecting 9 of the 20 pharmaceuticals and metabolites at all of the locations tested.[33] These plants provide drinking water for over 28 million Americans.

EPA funded a 2010 meta-analysis of 48 publications and found that 54 active pharmaceutical ingredients and 10 metabolites have been detected in treated drinking water.[34] The analysis notes that of the 64 substances that have been detected, only 36 have corroborative data from at least a second study.[35]

Studies Have Also Detected Occurrence of Pharmaceuticals in Treated Wastewater

In addition to source and treated drinking water, USGS and others have tested the effluent of wastewater treatment plants and animal feeding operations, two sources that are thought to be significant contributors of contaminants to streams and other sources of drinking water. Specifically:

- *Treated wastewater.* A 2005 study by USGS and EPA collected water samples upstream and downstream of wastewater treatment plants at

10 different locations totaling 40 sampling sites across the United States.[36] The agency tested for the presence of 110 chemicals, including industrial wastewater compounds and pharmaceuticals and related chemicals. Specifically, the study reported finding nonprescription pharmaceuticals in over 40 percent of the samples; prescription, nonantibiotic pharmaceuticals in over 30 percent of samples; and antibiotics in fewer than 10 percent of all samples. The study's results demonstrated an increase in the frequency of detection and concentration of most of the pharmaceuticals, and other chemical compounds, in the treatment plants' effluent as compared to water samples collected upstream of these plants; however, the chemical concentrations and occurrences decreased downstream from the treatment plants.

- *Animal feeding operations.* A study published in 2002 reported finding concentrations of antimicrobial agents in surface and ground water near large-scale poultry and swine farms, and concluded that animal waste likely acted as a source for antimicrobial residues in nearby water resources.[37] Specifically, the study noted that livestock receive antimicrobials both in therapeutic and nontherapeutic doses (i.e., in their feed), and that these compounds can be excreted into the environment.[38]
- *Pharmaceutical manufacturing facilities.* A 2010 USGS study of emerging contaminants in wastewater treatment plant effluents found that wastewater treatment plants that receive discharge from pharmaceutical manufacturing facilities had 10 to 1,000 times higher concentrations of pharmaceuticals (including opioids, muscle relaxants, and a barbiturate) than typically found in wastewater effluents. Maximum concentrations of some pharmaceuticals were in the part per million range.[39]

Research Has Not Determined the Human Health Effects of Pharmaceuticals in Drinking Water, but Some Research Raises Concerns

Research has not determined the human health effects of exposure to pharmaceuticals in drinking water. However, some research has demonstrated the potential impact to human health from exposure to some pharmaceuticals found in drinking water, such as EDCs and antibiotics.

Research Has Not Determined the Human Health Effects of Pharmaceuticals in Drinking Water

Uncertainty persists regarding whether pharmaceuticals in drinking water pose a risk to human health, and research has pointed to different conclusions.[40] For example, in its April 2008 testimony before the Senate Committee on Environment and Public Works,[41] the Pharmaceutical Research and Manufacturers of America, a trade association for the leading research-based pharmaceutical and biotechnology companies, cited a peer-reviewed study for which it provided financial support that concluded there was no demonstrable health risk to exposure to 26 pharmaceuticals detected by USGS in one of its national reconnaissance studies.[42,43] The study reached its conclusions by comparing an estimate of human exposure from drinking water and/or ingesting fish for each pharmaceutical to the acceptable daily intake (ADI) for that pharmaceutical. ADI is an estimate of the daily amount of pharmaceuticals that can be ingested by a healthy adult of normal weight and that should not result in an adverse health effect. In this instance, the ADI was derived from data developed by pharmaceutical manufacturers when testing the effectiveness and safety of a therapeutic dose of the pharmaceutical.

Other research has emphasized the absence of data and lack of knowledge regarding the health effects of pharmaceuticals in the environment. For example, research funded by EPA notes that risk assessments based on benchmarks such as ADIs generally conclude that there is negligible risk from exposure to pharmaceuticals through drinking water but that benchmark levels such as ADI are orders of magnitude higher than the exposure levels and may not take into account less obvious, nontherapeutic effects.[44,45] This research notes that despite the lack of empirical data linking pharmaceuticals in drinking water to adverse human health effects, the issue remains one of interest because of the unanswered questions concerning low-dose exposure to contaminants of emerging concern, including but not limited to pharmaceuticals. Some of the most significant unanswered questions identified in the research are:

- exposure to pharmaceuticals, including for sensitive subpopulations[46] such as children and in utero exposure?
- What are the effects of mixtures of pharmaceuticals, both additive and interactive?
- How do pharmaceuticals interact with the many other contaminants—both man-made and naturally occurring—that may be present in drinking water?

- Are there transgenerational effects (i.e., present in successive generations)?

Some Research Has Raised Concerns about EDCs and Antibiotics

The human health effects of pharmaceuticals in drinking water have not been conclusively shown, but research showing an impact on aquatic life raises concerns about two classes of pharmaceuticals—EDCs and antibiotics. Some of the concern about EDCs in drinking water stem from studies that have documented the abnormalities associated with aquatic life exposed to EDCs in rivers and lakes. Specifically, scientists have expressed concern because of both the significance of the abnormalities and the effects of contaminants on animals, which can be indicative of similar effects on humans. For example:

- A 2007 study reported that 75 percent of male smallmouth bass in certain areas of the South Branch of the Potomac River basin had ovarian tissue in their gonads.[47] The study concluded that a combination of EDCs was likely to have caused the feminization of the male fish. Although the authors note that the actual EDCs responsible for the abnormalities could not be determined, they suggest that a combination of contaminants could be the cause and noted that the additive effects of many EDCs have been demonstrated even when each compound present is below the threshold of detectable effects.
  The authors further noted that reproductive abnormalities in fish are frequently associated with human wastewater effluent, which contains synthetic estrogens found in birth control and hormone replacement medications.
- In another 2007 study by EPA and the Canadian government, researchers reported conducting a 7-year whole-lake experiment to test the effects on fathead minnows of chronic exposure to a synthetic estrogen used in some birth control pills.[48] The researchers reported a collapse in the population of fathead minnows in the experimental lake and concluded that the results from the study demonstrate that continued introduction of estrogens and estrogen mimics to the aquatic environment through municipal wastewaters could decrease the reproductive success and sustainability of fish populations.
- According to a 2004 research study, fish exposed to effluent from a cattle feedlot in Nebraska experienced reproductive abnormalities,

including reduced testes size in male fish and a lower level of estrogen in female fish.[49] The study reported the use of androgens in growth implants in the feedlot as one possible cause of the abnormalities.

Not all EDCs found in drinking water, however, are pharmaceuticals. Other contaminants, such as industrial chemicals and products, as well as naturally occurring hormones found in plants and excreted by different species, can also act as EDCs. Because other chemicals have also been shown to have potential endocrine-disrupting effects, the extent to which pharmaceutical EDCs contribute to detected abnormalities is unclear. For example, bisphenol A (BPA), a nonpharmaceutical EDC, is used to make polycarbonate plastics that are used in products such as compact disks, baby bottles, plastic dinnerware, eyeglass lenses, and toys. In its paper reporting 2003-2004 National Health and Nutrition Examination Survey findings, the Centers for Disease Control and Prevention found BPA in more than 90 percent of the urine samples representative of the U.S. population 6 years of age and older.[50] Another commonly occurring nonpharmaceutical EDC is atrazine, the most commonly used herbicide in the United States. In a 2003 study, scientists established a probable chain of causation between exposure to small concentrations of atrazine and the formation of female reproductive organs in frog testes.[51]

A second class of pharmaceuticals that has raised concern about the potential for health effects is antibiotics. In addition, some scientists are concerned about antimicrobial resistance resulting from interactions among chemicals, genes, microbes, animals, and humans in the environment. For example, some studies have demonstrated that bacteria exposed to pharmaceutical antibiotics and other antimicrobial agents in the environment have increased resistance to pharmaceutical antibiotics. However, the studies do not identify the extent to which pharmaceuticals or other antimicrobial agents contribute to these resistant bacteria. For example, triclosan and triclocarban, which are antimicrobials found in antiseptics, can contribute to antimicrobial resistance.[52] We recently issued a report that, among other issues, discusses scientific evidence supporting the association between antibiotic occurrence in the environment and an increase in resistance among bacteria.[53]

In addition to EDCs and antibiotics, other classes of pharmaceuticals have been found in drinking water and garnered scientific attention. Examples include chemotherapy drugs and selective serotonin reuptake inhibitors, which are a class of pharmaceuticals used to treat depression.

THE UNITED STATES AND OTHER COUNTRIES ARE ENGAGED IN TAKE-BACK PROGRAMS TO HELP PREVENT PHARMACEUTICALS FROM REACHING DRINKING WATER

Some states and local governments, as well as DEA, have taken actions to reduce the extent to which pharmaceuticals occur in drinking water— primarily through take-back programs to properly dispose of pharmaceuticals. These efforts are often tied to efforts to reduce drug abuse or accidental poisoning by removing expired medicines from the home. Through outreach and education on proper drug disposal, EPA has also taken steps to reduce the introduction of hazardous pharmaceutical waste into water supplies. Other countries—including Sweden and Australia—have undertaken additional efforts to reduce the occurrence of pharmaceuticals in drinking water.[54]

Consumer Drug Take-Back Programs Are Found Across the United States, and Congress and Others Are Considering Additional Efforts

Federal agencies do not have comprehensive data on the number of take-back programs across the United States, but EPA and the Product Stewardship Institute, Inc. collectively identified 25 states that have had one or more take-back programs.[55,56] In addition, DEA has held two nationwide take-back programs—in September 2010 and April 2011— and a third is planned for October 29, 2011.

Take-back programs are organized by a wide variety of stakeholders, including environmental groups, those with interests in preventing prescription drug abuse, and government entities (app. II provides federal guidelines on the proper disposal of pharmaceuticals). According to experts and program organizers we interviewed, the goals for implementing these programs include preventing drug abuse and accidental poisoning, as well as preventing unused pharmaceuticals from entering the environment. Pharmaceuticals collected through take-back programs are incinerated.

Through a survey of the literature and interviews with experts, we determined that take-back programs generally fall into one of three broad categories: (1) ongoing, (2) one-time, and (3) mail-back. To illustrate the three categories, we selected five take-back programs to review more closely. Figure 4 describes these five programs.

Program type	State	Coverage	Program name	Collection
Ongoing	Utah 	Statewide	Proper Medication Disposal	5,625 lbs (June 2009 to June 2010)
	Washington 	6 counties	PH:ARM pilot - (Pharmaceuticals from Households: A Return Mechanism)	15,134 lbs (June 2009 to June 2010)
One-time	California 	San Francisco Bay Area	Bay Area Pollution Prevention Group Safe Medicine Disposal Days	3,634 lbs (May 13-21, 2006)
	Texas 	Amarillo, Canyon, and their surrounding communities	Medication Cleanout™	1,947 lbs (September 12, 2009; March 27, 2010; June 9, 2010)
Mail-back	Maine 	Statewide	Safe Medicine Disposal for ME	2,373 lbs (May 2008 to October 2009)

Sources: GAO; art (Art Explosion).

Note: Data on the volume of pharmaceuticals collected generally include packaging.

Figure 4. Five Take-Back Programs GAO Identified.

As the figure shows, the following two programs are ongoing:

- *Utah's Proper Medication Disposal Program.* Consumers can leave unused pharmaceuticals in drop boxes at participating law enforcement agencies. The program collected over 5,600 pounds of pharmaceuticals, including packaging, from June 2009 to June 2010. It received $70,000 in grants from EPA and the Utah Department of Environmental Quality. The program costs, not including in-kind donations, were $40,000 from May 2007 to June 2010. According to program representatives, the program will seek additional grants to continue its efforts once it has spent the money from its current grants.
- *Washington State's PH:ARM Pilot (Pharmaceuticals from Households: A Return Mechanism).* PH:ARM began as a pilot project in 2006 with over 37 participating pharmacies in six counties. Consumers drop off their unused pharmaceuticals in secure drop

boxes at pharmacies. From October 2006 to October 2008, the program collected over 15,000 pounds of pharmaceuticals, including packaging, at a cost of approximately $170,000. According to program representatives, grant funding for the initial pilot project has ended, but the pharmacies have chosen to continue to collect unused pharmaceuticals on their own. Legislation proposed in the state legislature would have required pharmaceuticals manufacturers to pay for take-back programs in the state; however, the legislation failed a state senate vote in 2011.

We also identified one-time take-back events. These events are often organized by local communities and operate for a day, several days, or several weeks. For example:

- *Bay Area Pollution Prevention Group.* This group, a consortium of 43 wastewater agencies in the San Francisco Bay Area, piloted a week-long take-back program called "Safe Medicine Disposal Days" in May 2006. Consumers were invited to drop off pharmaceutical waste at 39 locations, including pharmacies, law enforcement offices, household hazardous waste facilities, and senior and civic centers. Over the course of the event, more than 1,500 residents disposed of over 3,600 pounds of pharmaceuticals. The event cost around $180,000, including administrative costs, and was funded by local agencies, cities, counties, and wastewater treatment plants.
- *Amarillo and Canyon, Texas, "Medication CleanoutTM" (MCO) pro*gram. Three 1-day events were conducted between September2009 and July 2010. These events were organized and funded by the Texas Panhandle Poison Center of Texas Tech University Health Sciences Center School of Pharmacy, the Amarillo Independent School District's *Safe Schools Healthy Students* program, and the Amarillo Police Department. Medication Cleanout provided consumers with drive-through drop-off points in order to return their unused pharmaceuticals without leaving their cars. The cost for the September 2009 event—the only date for which cost data are available—was approximately $44,000, and organizers reported that approximately 1,900 pounds of pharmaceuticals, including some packaging, were returned for all three events. Program organizers indicated that similar 1-day, drive-through events would be planned for the future.

Mail-back programs allow consumers to use the Postal Service to dispose of unused pharmaceuticals. For example, in 2008, Maine implemented a 2-year mail-back pilot program—called "Safe Medicine Disposal for ME." The program distributed postage-paid return envelopes to pharmacies and health and social service agencies across the state to be given to consumers. The envelopes contained instructions for how to properly return the pharmaceuticals, including how to remove personally identifying information from prescription bottles before mailing the unused pharmaceuticals. The pharmaceuticals were sent to the Maine Drug Enforcement Agency for proper disposal. Between May 2008 and October 2009, the program collected more than 2,600 pounds of pharmaceuticals, including packaging. Organizers reported that some of the prescriptions returned were over 20 years old. The program was initially funded with a $150,000 EPA grant and has since received $150,000 from the Fund for Healthy Maine that will allow the program to operate into 2011. Program organizers stated that their main goals for implementing the program were to prevent poisonings and drug abuse, but that 77 percent of respondents to a survey included with the envelopes distributed by the program reported that they participated because they were concerned about the environment.

According to DEA, its two nationwide take-back events—in September 2010 and April 2011—collected more than 300 tons of pharmaceuticals at thousands of sites across the country.

Although the U.S. take-back programs differ in how they are implemented, organizers of the events have faced similar challenges. For example, according to experts and organizers of the take-back programs we spoke with, these programs have been hampered by legal restrictions and limited funding, although the legal restrictions are being addressed. These experts and organizers told us that collecting controlled substances was resource intensive because, until recently, according to DEA the Controlled Substances Act made it was unlawful for the recipient of a controlled substance to give that substance to anyone other than law enforcement, even for the purposes of disposal.[57] Thus, consumers were prohibited from returning unused controlled substances to their pharmacy or doctor. Any take-back program that intended to collect controlled substances had to arrange for law enforcement to receive the unused controlled substances and maintain custody of them until they were destroyed.

However, in October 2010 the Secure and Responsible Drug Disposal Act was enacted amending the Controlled Substances Act. The act gives DEA the authority to issue regulations allowing communities and others to establish secure disposal programs for unused controlled substances. It also authorizes DEA to permit long-term care facilities to dispose of controlled substances on behalf of consumers who no longer need them. According to the Deputy Assistant Director of DEA's Office of Diversion Control, DEA strongly supported this legislation and anticipates issuing a notice of proposed rulemaking in the fall of 2011.

According to experts and program organizers, take-back programs are also hampered by limited funding. Programs use a combination of in-kind contributions, volunteer time, grants, and local funding sources to pay for their programs. For example, between 2004 and 2008, EPA awarded 25 grants—totaling $926,972—to support take-back programs; these grants ranged from approximately $10,000 to $150,000. In addition, at least one state has previously proposed legislation that would require pharmaceutical manufacturers to fund take-back programs. As of March 2011, no such state legislation had been enacted.

Other Countries Also Support Take-Back Programs

In 2004, the European Union (EU) issued a directive to its member states to, among other things, ensure that appropriate collection systems are in place for medicinal products that are unused or have expired in light of the potential risks presented by these pharmaceuticals for the environment."[58] Three years later, in 2007, the European Federation of Pharmaceutical Industry Associations surveyed 27 EU member states on their implementation of programs to collect unused pharmaceuticals. Of the 22 national pharmaceutical associations responding to the survey, 19 reported they had a pharmaceutical waste collection program, and most of these 19 associations reported that the programs operate nationwide. In 6 of the 19 programs, the pharmaceutical industry funds all costs associated with collecting and destroying unused pharmaceuticals.

Sweden is an example of an EU country that has taken additional steps to reduce the occurrence of pharmaceuticals in drinking water. Sweden's efforts are supported by its government; pharmacies (most of which are publicly owned) are now obligated to take back all unused or expired pharmaceuticals and safely incinerate them. In 2009, 1,128 tons of pharmaceuticals, including packaging, were returned and destroyed. Sweden has also taken the following actions:

- *Classifying pharmaceuticals according to how toxic they would be if they were released into the environment.* According to a Swedish official, in 2004, officials from pharmaceutical producers and Sweden's health care system created an environmental classification system for pharmaceuticals to provide doctors and patients with information about the environmental effects of pharmaceuticals. Sweden developed this system by using risk and hazard data submitted by pharmaceutical manufacturers on their products. These data were then evaluated by an independent consulting firm, which provided an approval or disapproval for the proposed risk and hazard levels. The pharmaceuticals' risk and hazard determinations used the following criteria: biodegradability, potential to accumulate in the body, and toxicity to aquatic organisms. Individual jurisdictions throughout Sweden then used these results to compile lists of pharmaceuticals recommended for specific ailments, and doctors may consider these lists when prescribing pharmaceuticals. In addition, at least one pharmaceutical company has indicated that it is pursuing initiatives to produce less toxic and more environmentally friendly pharmaceuticals.
- *Encouraging initial prescriptions in smaller amounts.* According to data from Sweden, in 2005 and 2006, nearly 40 percent of the pharmaceuticals collected were unopened, and the remaining packages were still nearly two-thirds full, suggesting that patients may be buying more pharmaceuticals than they need. As a result, the public providers of healthcare encourage doctors to prescribe smaller initial prescriptions so that patients and their physician can determine if the pharmaceutical will work for the patient. This practice may reduce the amount of pharmaceuticals that are disposed of when patients switch to different pharmaceuticals.

According to one knowledgeable Swedish official, Sweden adopted these policies—even though there is no scientific evidence that the occurrence of pharmaceuticals in the environment is affecting human health—as a result of its adherence to the "precautionary principle." This principle states that action should be taken without waiting for the certainty of causation when an appropriate level of scientific evidence suggests an association between hazardous environmental exposures and ill health. According to the principle, action should be taken preventively because definitive knowledge about causation might take decades of further research.

Outside of the EU, Australia has a national take-back program—"Return Unwanted Medicines" (RUM). RUM is a national, government-financed program that allows consumers to return unwanted or expired pharmaceuticals to participating pharmacies. Educational materials from the RUM program instruct consumers that they should not dispose of pharmaceuticals in the trash, in the toilet, or in the sink. According to RUM data from July 2009 through June 2010, the RUM project collected 1,075,957 pounds of pharmaceutical waste, including packaging, that might otherwise have been disposed of through wastewater or in the trash and risk contaminating the environment. A program representative stated that RUM has been an integral component of Australia's efforts to advise consumers on all aspects of pharmaceutical consumption and disposal.

DATA GAPS MAKE IT DIFFICULT FOR EPA TO IDENTIFY PHARMACEUTICALS FOR REGULATION; SUSTAINED COLLABORATION MAY HELP EPA ADDRESS SUCH DIFFICULTIES

EPA faces challenges in obtaining sufficient occurrence and health effects data to support analyses and decisions about which pharmaceuticals to include on the Contaminant Candidate List as well as to make regulatory determination decisions. EPA is collaborating with other agencies on research to help obtain these data for use in developing future candidate lists, but these efforts are largely informal and EPA has not established a formal mechanism to sustain these collaborative efforts. We previously reported key practices for enhancing and sustaining collaboration among federal agencies that may be an option to help institutionalize an approach for conducting research that leverages resources among the agencies.[59] We recommended that the Director of the Office of Management and Budget continue to encourage interagency collaboration by among other things, promoting and collaboration practices identified in GAO's report; the Office of Management and Budget agreed with the recommendation.

EPA Faces Challenges Because of Gaps in Occurrence and Health Effects Data

EPA faces significant data gaps concerning both the occurrence and health effects of pharmaceuticals. Sufficient occurrence and health effects data are critical for EPA to assess pharmaceuticals for possible regulatory determinations under the criteria established by SDWA. The difficulties EPA experienced in evaluating pharmaceuticals to include on its most recent Contaminant Candidate List, in 2009, illustrate the challenges EPA faces in obtaining these data.

Occurrence Data Were Limited for Pharmaceuticals Considered for the Most Recent Contaminant Candidate List

To evaluate pharmaceuticals for inclusion on its 2009 Contaminant Candidate List, EPA identified two general types of occurrence data: first, data on the actual detection of pharmaceuticals in source and treated drinking water, and second, data on environmental releases and production volumes of pharmaceuticals developed by industry and government.[60]

Source and treated drinking water: EPA occurrence data on pharmaceuticals detected in untreated source water came from USGS's national reconnaissance study on surface water and related efforts. These efforts provided data on 123 contaminants, including pharmaceuticals. The data contain measurements of contaminants in water but the data were from sample sites often chosen because they were predicted to be the most likely place that pharmaceuticals and other emerging contaminants would enter the environment (e.g., downstream from wastewater treatment plants). The sample sites are not statistically representative of average conditions across the nation. However, the sites were geographically distributed and included a mix of characteristics that were intended to provide a basic understanding of whether pharmaceuticals and other contaminants are in the nation's waterways. According to EPA, the most relevant occurrence data are for treated drinking water, but these data are often not available. EPA told us it evaluated the available studies from the scientific literature that included occurrence data for pharmaceuticals from treated drinking water, but there were only a limited number of studies available and the majority of these studies only sampled a limited number of drinking water systems. Thus, to identify pharmaceuticals for inclusion on the most recent candidate list, EPA instead relied on data on untreated source water. Most Americans consume treated drinking water.

Environmental release and production volumes: EPA also obtained occurrence data on pharmaceuticals from the Toxics Release Inventory and the High Production Volume Chemical List. The Toxics Release Inventory contains industry- and government-reported information on chemical releases into the environment—air, land, and water; the High Production Volume Chemical List contains production volume information for chemicals manufactured or imported into the United States in quantities greater than certain threshold amounts. However, EPA considered these data sources to provide less meaningful information on a chemical's potential to occur in drinking water than sources that actually detect the presence of chemicals in the environment, such as the USGS data that it did use.

For the 12 pharmaceuticals that it included on its 2009 Contaminant Candidate List, EPA reported it does not have comprehensive occurrence data for treated drinking water for any of them and does not have an analytic method suitable for conducting national drinking water studies for 7 of them. For the remaining 5 pharmaceuticals, EPA reports that it has or is developing a suitable analytic method.[61]

Health Effects Data Were Limited for Pharmaceuticals Considered for the Most Recent Contaminant Candidate List

According to the *Federal Register* notice for the draft 2009 Contaminant Candidate List,[62] the primary source of health effects information on pharmaceuticals in drinking water was the FDA database on maximum recommended daily doses. This FDA database includes the recommended doses for the "average adult patient" for over 1,200 pharmaceuticals and is based on human clinical trials of daily exposure, usually for 3 to 12 months. The maximum recommended daily dose is an estimated upper dose beyond which a pharmaceutical is not more effective and/or adverse effects begin to outweigh beneficial effects. However, according to EPA-sponsored research,[63] extrapolating health effects data from data on the therapeutic doses of individual pharmaceuticals does not address, among other issues, the following two areas of concern about pharmaceuticals in drinking water: the health effects of (1) long-term, low-dose exposure to pharmaceuticals and (2) exposure to mixtures of pharmaceuticals.

- *Effects of long-term, low-dose exposure to pharmaceuticals.* According to the EPA-sponsored research, the health effects of long-term, low-dose exposure to a pharmaceutical may not be predictable by extrapolating from an observed effect of shorter-term exposure to much higher concentration of that pharmaceutical. The research indicates that further complications arise when trying to predict the effects of exposure on sensitive sub-populations. For example, a child in the age group between birth and 1 month might be particularly sensitive to a contaminant during this life stage, during which the child experiences rapid growth, weight gain, and immature immune system function, among other characteristics, which can influence a child's susceptibility to a particular chemical.[64]

- *Effects of exposure to mixtures of pharmaceuticals.* Also according to the EPA-sponsored research, the simultaneous exposure to multiple pharmaceuticals could result in an additive or interactive effect. In particular, studies on occurrence data have found more than one contaminant in a single water sample. For example, the USGS national reconnaissance study on surface water that EPA used to identify contaminants for the most recent candidate list found that there was a median of 7, and as many as 38, of the tested contaminants in a given sample.[65]

For the 12 pharmaceuticals that it included on the most recent candidate list, EPA reported that it has substantial data needs on health effects for 8 of them. For the remaining 4 pharmaceuticals, EPA reports that information exists or there is an ongoing assessment. Furthermore, as we recently reported, EPA has not identified the drinking water contaminants of greatest public health concern. In many cases, gathering sufficient data to make a regulatory determination has taken EPA more than 10 years, and obtaining data on other contaminants on the current list may well take decades.[66] We made recommendations regarding the need for EPA to develop criteria to identify contaminants that pose the greatest health concern and a process to obtain data to support regulatory determinations; EPA did not agree to adopt these recommendations and generally took the position that no further steps are needed.

EPA is collaborating with other federal agencies to collect occurrence and health effects data on pharmaceuticals and other contaminants that could support decisions about which contaminants to include on future candidate lists as well as regulatory determinations. As the following examples demonstrate, collaboration is helping EPA leverage the resources and expertise of other agencies to obtain results that may have been more difficult for it to achieve on its own.

EPA Is Collaborating with Other Agencies to Obtain Needed Occurrence and Health Effect Data on Pharmaceuticals, but It Has not Established a Formal Mechanism to Sustain Future Collaboration

- EPA and USGS are jointly developing occurrence data for over 230 contaminants, more than half of which are pharmaceuticals, in a study designed to provide EPA with data for future candidate lists. The agencies' joint study will sample treated drinking water and source water in about 25 drinking water treatment plants across the nation. These plants were selected because they draw water from streams, lakes, reservoirs, or ground-water aquifers affected by a variety of waste sources (e.g., municipal waste, septic systems, livestock production). EPA is providing expertise to analyze micro-organisms, and has experience with drinking water treatment facilities and their design. USGS is providing its expertise in the logistics of operating a nationwide water sampling project. Both agencies have expertise in detecting low concentrations of pharmaceuticals and other contaminants of emerging concern. The study is expected to conclude in September 2012.
- EPA is working with FDA to develop a methodology to more efficiently assess the health effects of pharmaceuticals in drinking water by addressing groups of related pharmaceuticals, such as selective serotonin reuptake inhibitors, instead of individual pharmaceuticals. FDA is providing health effects data, and EPA plans to use the methodology to support decisions about which pharmaceuticals to include on future candidate lists. This effort is part of a larger EPA initiative to better implement SDWA by focusing on assessing risk from exposure to groups of contaminants instead of individual contaminants.

According to EPA officials, there is no formal mechanism, such as a long-term strategy or formal agreement, to manage and sustain these collaborative efforts. Agency officials and former members of the PiE workgroup told us that interagency efforts such as those described above are the result of informal collaborative relationships among agency personnel, particularly those fostered by the PiE workgroup. As one official from EPA's Office of Water noted, the current interagency collaboration is "ad hoc." In 2008 and 2010, we reported that by using informal coordination mechanisms, agencies may rely on relationships with individual officials to ensure effective collaboration, but these informal relationships could end if the responsible staff are not available to continue the efforts.[67] We recommended that those agencies develop clear guidance for interagency planning efforts in the 2008 report, and that roles and responsibilities be identified to support collaboration in the 2010 report; the agencies generally agreed with these recommendations.

The purpose of the PiE workgroup was to identify and prioritize research needed to better understand the risk from pharmaceuticals in the environment and to recommend areas for federal collaboration to address those priorities.[68] Its draft report was neither approved by NSTC nor publicly released.[69] According to OSTP officials, the draft report was not approved or released because the workgroup did not address OSTP's concerns, including that the report did not specifically outline how agencies would coordinate research and other long-term activities identified in the draft report once the workgroup expired. For example, OSTP officials stated that the draft report did not clarify collaborating agencies' roles and responsibilities by identifying which agencies are best positioned to address specific issues identified in the report and which existing or new programs would be most appropriate for addressing these issues. OSTP officials told us that providing this additional information was consistent with the purpose of the workgroup. The workgroup co-chairs told us that OSTP did not present the workgroup with its written concerns until June 2010, about a year after the draft report was approved by the Subcommittee on Toxics and Risk, and after the workgroup had expired. According to the co-chairs, addressing OSTP's concerns would have required the workgroup to update the scientific data included in the draft report and would have required the workgroup to provide additional information regarding agencies' roles and responsibilities that was beyond the purpose of the workgroup. Thus, the draft report was never finalized although, according to the co-chairs, the interagency activities begun by the workgroup continued.

Key Practices Can Enhance and Sustain Coordination

In an October 2005 report, we identified key practices for enhancing and sustaining collaboration among federal agencies. Three of these practices may help clarify how EPA and other agencies could coordinate their research efforts.[70]

- *Establish roles and responsibilities*: We reported that collaborating agencies should work together to define and agree on their respective roles and responsibilities, including how the collaborative effort will be led. In doing so, agencies can clarify who will do what, organize their joint and individual efforts, and facilitate decision making.
- *Leverage resources*: We reported that collaborating agencies should identify the human, information technology, physical, and financial resources needed to initiate or sustain their collaborative effort. Collaborating agencies bring different levels of resources and capacities to the effort. By assessing their relative strengths and limitations, collaborating agencies can look for opportunities to address resource needs by leveraging each other's resources, thus obtaining additional benefits that would not be available if they were working separately.
- *Establish mechanisms for monitoring, evaluating, and periodically reporting results of the collaborative research efforts*: We reported that federal agencies engaged in collaborative efforts need to create the means to monitor and evaluate their efforts to enable them to identify areas for improvement. Reporting on these activities can help key decision makers within the agencies to obtain feedback for improving both policy and operational effectiveness.

CONCLUSION

There are basic questions about the potential health risks from exposure to pharmaceuticals in the nation's drinking water. Other contaminants also have been detected in drinking water including personal care products, and chemicals used in industry and agriculture, including some that may act as EDCs. Some of these other contaminants may work in tandem with pharmaceuticals to affect human health through additive or interactive effects. Also of concern to government scientists are the health effects of long-term low-dose exposure to pharmaceuticals and exposure to mixtures of pharmaceuticals.

Since the 1996 amendments to SDWA, EPA has been required to publish a list of currently unregulated contaminants including pharmaceuticals that may require regulation in drinking water, and to make determinations on whether or not to regulate at least 5 of the contaminants on the list every 5 years. In 2009, EPA issued its third Contaminant Candidate List, which consists of 116 contaminants, 12 of which are pharmaceuticals. However, EPA continues to experience difficulty obtaining sufficient occurrence and health effects data for making determinations on (1) which contaminants present the greatest public health concern to include on the list and (2) whether or not to regulate any of the contaminants on the candidate list. It will continue to be difficult for EPA to prioritize contaminants on the candidate list without the necessary information on health effects and occurrence to determine the contaminants that present the greatest public health concern. In many cases, gathering sufficient data to address contaminants awaiting determinations has taken EPA more than 10 years, and obtaining data on other contaminants on the current list may well take decades. To collect occurrence and health effects data on pharmaceuticals and other contaminants that could support decisions about which contaminants to include on future candidate lists, EPA is collaborating informally with USGS and FDA, but does not have a formal mechanism for sustaining such collaboration in the future. Furthermore, the PiE workgroup, which pulled together the scientific expertise from eight federal agencies, has expired and its draft report was neither finalized nor released.

However, neither EPA's informal collaboration efforts nor the strategy proposed by the PiE workgroup details how agencies could coordinate their future interagency collaboration efforts. We have previously reported on key practices for enhancing and sustaining interagency collaboration efforts, such as (1) establishing roles and responsibilities, including how the collaborative effort will be led; (2) identifying the expertise and other resources that each agency can bring to bear on the issue, and (3) establishing a process for monitoring, evaluating, and reporting to the public the results of the collaborative research efforts.

Recommendation for Executive Action

To collect the pharmaceutical occurrence and health effects data necessary to better implement SDWA, and to address the broader issue of pharmaceuticals and their relationship to other contaminants in the nation's waterways, we are making the following recommendation to the Administrator of EPA:

- Establish a workgroup or other formal mechanism that includes the relevant federal agencies to collaborate and coordinate research on pharmaceuticals and, as appropriate, other contaminants in drinking water that present the greatest public health concern. In establishing this mechanism, EPA should: (1) define roles and responsibilities, including how the collaborative effort will be led; (2) identify the expertise and other resources that each agency can bring to bear on the issue; and (3) develop a process for monitoring, evaluating, and reporting to the public the results of the collaborative research efforts.

AGENCY COMMENTS AND OUR EVALUATION

We provided a draft of this report to EPA, the Department of the Interior (DOI), the Department of Health and Human Services (HHS), OSTP, and the Department of Justice (DOJ) for review and comment.

In written comments, EPA agreed with our findings and recommendation and noted that the extent of interagency collaboration may be dependent upon available resources. EPA also provided clarifying language regarding the responsibilities, accomplishments, and activities of the PiE workgroup which, according to EPA, reflects clarification provided by the PiE workgroup co-chairs. We modified our draft accordingly. EPA's comments are reprinted in appendix III. EPA also provided technical clarifications and comments, which we incorporated as appropriate.

DOI also provided written comments on a draft of this report and stated that it generally agrees with the findings and recommendation in the report. Additionally, DOI provided clarifying language regarding the PiE workgroup. DOI's comments are reprinted in appendix IV. Additionally, USGS, an agency within DOI, provided technical clarifications and comments, which we incorporated as appropriate.

DOJ, HHS, and OSTP did not provide written comments but provided technical clarifications and comments, which we incorporated as appropriate.

David C. Trimble
Director, Natural Resources and Environment

Appendix I: Objectives, Scope, and Methodology

The objectives of this study were to (1) provide information on the extent to which pharmaceuticals occur in drinking water and the effects, if any, that their occurrence has on human health; (2) describe the approaches taken in the United States and in other countries to reduce the extent to which pharmaceuticals occur in drinking water; and (3) identify challenges, if any, that the Environmental Protection Agency (EPA) faces in determining whether any pharmaceuticals should be regulated under the Safe Drinking Water Act (SDWA), actions EPA is taking to address these challenges, and options for addressing such challenges in the future.

To identify the extent to which pharmaceuticals occur in drinking water, we reviewed federal and peer-reviewed reports, including (1) studies by the U.S. Geological Survey (USGS), (2) articles in peer-reviewed journals by federal scientists and others, and (3) the Pharmaceuticals in the Environment (PiE) workgroup's draft report. We also selected a nonprobability sample of scientific studies to review in our report. The data from these studies are not generalizable beyond the scope of these studies. We selected these studies on the basis of certain criteria, including the source of the study (e.g., a peer-reviewed journal); the geographic scope of the study; and whether the study focused on source water, treated drinking water, or wastewater. We also discussed the subject with scientists at USGS and other federal agencies as well as with representatives from academia, trade associations, the environmental community, and the pharmaceutical industry.

To identify the effects, if any, that the occurrence of pharmaceuticals in drinking water has on human health, we also reviewed federal and peer-reviewed reports, including articles in peer-reviewed journals by federal scientists and others; and the PiE workgroup's draft report. We discussed the subject with federal scientists and representatives from academia, the environmental community, and the pharmaceutical industry. We also attended an October 2009 academic conference on hormones and related compounds in the environment that was hosted by Tulane University.

To describe the approaches taken in the United States to reduce the extent to which pharmaceuticals occur in drinking water; we reviewed literature and spoke with officials from federal agencies including the Drug Enforcement Administration (DEA), EPA, and the Food and Drug Administration (FDA), as well as experts from academia, industry and nonprofit organizations that have ongoing work addressing pharmaceuticals in the environment; from these efforts, we identified consumer take-back programs as the primary approach to reducing occurrence. We also determined that take-back programs could be grouped into three broad categories based on common characteristics— mail back, one-time, and ongoing. We selected a nonprobability sample of five programs to represent the three categories. The information from these programs is not generalizable to all take-back programs. We selected the programs because they provided geographic diversity and exemplified certain characteristics. For example, we selected one program, in part, because it was pharmacy-based. We did not attempt to evaluate the programs. We collected information on each program through a survey, follow-up interviews, and, where appropriate, additional documentation. To describe approaches taken by other countries to reduce the extent to which pharmaceuticals occur in drinking water, we chose to describe efforts in Sweden and Australia. We selected Sweden because it is undertaking a variety of stewardship activities. We selected Australia because it has a nationwide take-back program. We obtained information on each country's efforts though interviews with knowledgeable officials and, where appropriate, additional documentation.

To identify challenges, if any, that EPA faces in determining whether any pharmaceuticals should be regulated under SDWA, actions EPA is taking to address these challenges, and options for addressing such challenges in the future, we reviewed agency documents and interviewed relevant agency officials. Specifically, to identify challenges, we reviewed EPA's documentation on the process it used to develop the 2009 Contaminant Candidate List under the authority of SDWA. We also reviewed some of the sources of data that EPA relied upon to identify pharmaceuticals for inclusion on the candidate list. To identify actions that EPA is undertaking to address challenges we identified, we interviewed agency officials from EPA, FDA, and USGS and, where appropriate, obtained and reviewed additional documentation. To identify options to address these challenges in the future, we obtained and reviewed a 2009 draft report produced by the PiE workgroup. We also interviewed several of the workgroup members, including the three co-chairs. We also reviewed our own work on practices that can help enhance and sustain collaboration among federal agencies.

We conducted this performance audit from January 2010 through August 2011 in accordance with generally accepted government auditing standards. Those standards require that we plan and perform the audit to obtain sufficient, appropriate evidence to provide a reasonable basis for our findings and conclusions based on our audit objectives. We believe that the evidence obtained provides a reasonable basis for our findings and conclusions based on our audit objectives.

Appendix II: Guidelines for Proper Disposal of Unused Pharmaceuticals

According to FDA and the White House Office of National Drug Control Policy, consumers are encouraged to properly dispose of unused pharmaceuticals to avoid harm to others. In general, consumers should not flush prescription pharmaceuticals down the toilet or sink drain unless the label or accompanying patient information specifically instructs consumers to do so. However, in some instances, it may be necessary to dispose of unused pharmaceuticals by flushing. For a list of pharmaceuticals that are recommended to be flushed, consumers should visit FDA's Web site.[1]

Several disposal options are available to consumers for prescription pharmaceuticals that are not specifically labeled to be flushed. For example, other than the pharmaceutical take-back programs presented in this report, programs such as household hazardous waste collection events, which collect pharmaceuticals at a central location, can provide consumers with proper disposal of unused pharmaceuticals.

Organizations such as the Product Stewardship Institute have information on such events across the nation.

In addition, FDA and the White House Office of National Drug Control Policy recommend that consumers consider the following steps to dispose of unused pharmaceuticals:

1. Take prescription pharmaceuticals out of their original containers.
2. Mix pharmaceuticals (do NOT crush tablets or capsules) with an undesirable substance, such as cat litter or used coffee grounds.
3. Place the mixture into a disposable container with a lid, such as an empty margarine tub, or into a sealable bag.
4. Conceal or remove any personal information, including Rx number, on the empty containers by covering it with black permanent marker or duct tape, or by scratching it off.
5. Place the sealed container with the mixture, and the empty pharmaceutical containers, in the trash.

Appendix III: Comments from the Environmental Protection Agency

United States Environmental Protection Agency
Office Of Water
Washington, D.C. 20460

July 22, 2011

Mr. David C. Trimble
Acting Director
Natural Resources and Environment
U.S. Government Accountability Office
Washington, D.C. 20548

Dear Mr. Trimble,

The U.S. Environmental Protection Agency (EPA) appreciates the opportunity to review and comment on the Government Accountability Office (GAO) draft report, "Action Needed to Sustain Agencies' Collaboration on Pharmaceuticals in Drinking Water" (Report Number GA0-11-346). EPA agrees with the recommendation in the draft report with the caveat that level of engagement is dependent upon available resources. Brief comments on the EPA's position regarding the recommendation are provided below. In addition, EPA is providing clarifying language for two specific places in the draft report regarding responsibilities, accomplishments, and activities of the Interagency Pharmaceuticals in the Environment (PiE) workgroup. This additional language reflects clarification provided by the PiE workgroup co-chairs. Other technical comments and clarifications on draft report language have been provided in a separate document.

GAO Recommendation

To collect the occurrence and health effects data necessary to better implement SDWA, and to address the broader issue of pharmaceuticals and their relationship to other contaminants in the nation's waterways, GAO made the following recommendation to the Administrator of EPA Establish a workgroup or other formal mechanism that includes the relevant federal agencies to collaborate and coordinate research on pharmaceuticals and, as appropriate, other contaminants in drinking water that present the greatest public health concern. In establishing this mechanism, the EPA should:

- Define roles and responsibilities, including how the collaborative effort will be led;
- Identify the expertise and other resources that each agency can bring to bear on the issue; and;
- Develop a process for monitoring, evaluating, and reporting to the public the results of the collaborative research efforts.

EPA Response to Recommendation

The EPA supports the establishment of an interagency workgroup or other formal mechanism to collaborate and coordinate research on pharmaceuticals and, as appropriate, other contaminants in drinking water that present the greatest public health concern. The EPA anticipates that the extent to which collaborative efforts are conducted may be dependent upon available resources.

Clarifying Language

The PiE workgroup co-chairs recommend that the following language should replace the paragraph that starts at the bottom of page 12 and ends on page 13, significant additions to the draft report language are underlined.

The PiE workgroup was established in 2006 by the Committee on Environment, Natural Resources, and Sustainability (CENRS), Toxics and Risk Subcommittee, an executive branch entity under the National Science and Technology Council (NSTC).[Footnote 13] NSTC is a council of cabinet-level officials chaired by the President and managed by the Director of Office of Science and Technology Policy (OSTP).[Footnote 14] The purpose of the workgroup was to identify and prioritize research needed to better understand the risk from pharmaceuticals in the environment and to recommend areas for federal collaboration to address those priorities.[Footnote 15] The workgroup, which was intended to be temporary, was staffed by scientists from eight federal agencies.[Footnote 16] The EPA, FDA, and USGS scientists served as co-chairs. In May 2009, the PiE workgroup produced a draft report *that was reviewed and approved by the Toxics and Risk Subcommittee and then conveyed to OSTP.* OSTP officials told us they had several significant concerns with the draft report and indicated that it would not be approved or publicly released until the workgroup addressed these concerns. The workgroup co-chairs told us that the requested changes *were shared with them after the workgroup had expired and, as a result, the document was not finalized. However, interagency collaborative efforts begun by the workgroup have continued.*

The PiE workgroup co-chairs recommend that the following language replace the paragraph that starts at the bottom of page 38 and ends on page 39; significant additions to the draft report language are underlined:

The purpose of the PiE workgroup was to identify and prioritize research needed to better understand the risk from pharmaceuticals in the environment and to recommend areas for federal collaboration to address those priorities.[Footnote 66] The draft report was neither approved by NSTC, nor publicly released. *According to the NSTC 2008 Handbook, NSTC documents must be cleared by the White House Co-Chair of the Committee (in this case, CENRS), or by the OSTP Director, or, in the event that neither is available, by the OSTP General Counsel.* According to OSTP officials, the draft did not specifically outline how agencies would coordinate research and other long-term activities identified in the draft report once the workgroup disbanded. For example, OSTP officials stated that the draft report did not clarify collaborating agencies' roles and responsibilities by identifying which agencies are best positioned to address specific issues identified in the report and which existing or new programs would be most appropriate for addressing these issues.

The workgroup co-chairs told us that OSTP did not provide the workgroup with its concerns until after the workgroup had expired. Nonetheless, the interagency activities begun by the workgroup were continued.

Once again, thank you for the opportunity to respond to this draft report.

Sincerely,
Nancy K. Stoner
Acting Assistant Administrator

Appendix IV: Comments from the Department of the Interior

United States Department of the Interior
Office Of The Secretary
Washington, D.C. 20240

July 21, 2011

Mr. David Trimble
Acting Director, Natural Resources and Environment
U.S. Government Accountability Office
441 G Street, N.W.
Washington, D.C. 20548

Dear Mr. Trimble,

The Department of the Interior has reviewed the draft GAO report entitled Environmental Health: Action Needed to Sustain Agencies' Collaboration on Pharmaceuticals in Drinking Water (GAO-11-346).

The Department of the Interior generally agrees with the findings and recommendation in the report. However, the description of the responsibilities, activities, and accomplishments of the Pharmaceuticals in the Environment (PiE) Working Group, established in 2006 by the Committee on Environment, Natural Resources, and Sustainability (CENRS), Toxics and Risk Subcommittee, an executive branch entity under the National Science and Technology Council (NSTC), should be clarified. Language is provided to make that clarification in two specific places in the draft report.

The following language should replace the paragraph that starts at the bottom of page 12 and ends on page 13; significant additions to the draft report language are underlined.

The PiE workgroup was established in 2006 by the Committee on Environment, Natural Resources, and Sustainability (CENRS), Toxics and Risk Subcommittee, an executive branch entity under the National Science and Technology Council (NSTC).[Footnote 13] NSTC is a council of cabinet-level officials chaired by the President and managed by the Director of Office of Science and Technology Policy (OSTP).[Footnote 14] The purpose of the workgroup was to identify and prioritize research needed to better understand the risk from pharmaceuticals in the environment and to recommend areas for federal collaboration to address those priorities.[Footnote 15] The workgroup, which was intended to be temporary, was staffed by scientists from eight federal agencies.[Footnote 16] EPA, FDA, and USGS scientists served as co-chairs. In May 2009, the PiE workgroup produced a draft report [begin underline] that was reviewed and approved by the Toxics and Risk Subcommittee and then conveyed to OSTP.[end underline] OSTP officials told us they had several significant concerns with the draft report and indicated that it would not be approved or publicly released until the workgroup addressed these concerns. The workgroup co-chairs told us that the requested changes [begin underline] were beyond the charge of the workgroup and would reflect a markedly different document. The workgroup expired in 2009 and the report was never finalized. However, interagency collaborative efforts begun by the workgroup have continued. [end underline]

We recommend that the following language replace the paragraph that starts at the bottom of page 38 and ends on page 39; significant additions to the draft report language are underlined.

The purpose of the PiE workgroup was to identify and prioritize research needed to better understand the risk from pharmaceuticals in the environment and to recommend areas for federal collaboration to address those priorities.[Footnote 66] [The] draft report was neither approved by NSTC, nor publicly released. [begin underline] According to the NSTC 2008 Handbook, NSTC documents must be cleared by the White House Co-Chair of the Committee (in this case. CENRS), or by the OSTP Director. or. in the event that neither is available, by the OSTP General Counsel. [end underline] According to OSTP officials, the draft did not specifically outline how agencies would coordinate research and other long-term activities identified in the draft report once the workgroup disbanded. For example, OSTP officials stated that the draft report did not clarify collaborating agencies' roles and responsibilities by identifying which agencies are best positioned to address specific issues identified in the report and which existing or new programs would be most appropriate for addressing these issues. [begin underline] The workgroup co-chairs told us that OSTP did not provide the workgroup with its concerns until June 2010, about a year after the draft report was approved by the Subcommittee on Toxics and Risk, and after the workgroup had expired. According to the co-chairs addressing OSTP's concerns would have required the workgroup to update the draft report with significant additional information, including agencies' roles and responsibilities, which they felt would be beyond the purpose of the workgroup. Thus, the workgroup was not reactivated by the Toxics and Risk Subcommittee: however, the interagency activities begun by the workgroup were continued.

We hope these comments will assist you in preparing the final report.

Sincerely,

Anne J. Castle:
Assistant Secretary for Water and Science

End Notes

[1] This report uses the term "drinking water" to refer to treated drinking water—the water that has been treated before it enters homes or businesses. We use the term "source water" to refer to the water in rivers, lakes, ground water, and other water bodies that may be the source of drinking water before treatment or that may be consumed without being treated. We also use the term "effluent" to refer to treated wastewater.

[2] This report uses the term "pharmaceutical" to refer to active pharmaceutical ingredients (API) and related chemicals such as metabolites. According to the Food and Drug Administration, an API is a substance that is intended to furnish pharmacological activity or other direct effect in the diagnosis, cure, mitigation, treatment, or prevention of disease, or to affect the structure or any function of the body. A metabolite is a substance that is the product of biological changes to a chemical.

[3] EDCs are chemical compounds found in some pharmaceuticals, food, and consumer products, or occur naturally in the environment. EDCs can interfere with the functioning and development of hormones in humans and animals and can produce adverse developmental, reproductive, neurological, and immune effects.

[4] The Administrator may also regulate contaminants not on the Contaminant Candidate List if the criteria for regulation are met, but to date has not done so.

[5] 76 Fed. Reg. 7762 (Feb. 11, 2011).

[6] 74 Fed. Reg. 51850 (Oct. 8, 2009).

[7] EPA used a different approach to determine which contaminants to include on the first and second candidate lists. See 73 Fed. Reg. 9628, 9631 (Feb. 21, 2008).

[8] For additional information on the process EPA used to develop the third Contaminant Candidate List and the contaminants on that list, see http://www.epa.gov/ogwdw/ccl/ccl3.html.

[9] GAO, Safe Drinking Water Act: EPA Should Improve Implementation of Requirements on Whether to Regulate Additional Contaminants, GAO-11-254 (Washington, D.C.: May 27, 2011).

[10] EPA Office of Water, Guidance Document: Best Management Practices for Unused Pharmaceuticals at Health Care Facilities, Draft (August 2010): table 1, 14-15. EPA, Proposed Rule, Amendment to the Universal Waste Rule: Addition of Pharmaceuticals, 73 Fed. Reg. 73520, 73522 (Dec. 2, 2008).

[11] EPA Office of Water, Guidance Document: Best Management Practices for Unused Pharmaceuticals at Health Care Facilities, Draft (August 2010): 18.

[12] EPA noted it has also funded and assisted in the development of a guidance document to help healthcare facilities develop and implement hazardous pharmaceutical waste management plans.

[13] 40 C.F.R § 131.11 (2011). Most water quality criteria are expressed as numeric or quantitative-parameters. For example, national recommended water quality criteria for toxic pollutants are numeric and specify the precise, measurable levels of particular chemicals or conditions allowable in a water body. When numerical criteria cannot be established, such as when pollutants cannot be precisely measured, narrative criteria are used to express a parameter in a qualitative form.

[14] Effluent refers to wastewater discharged into the environment. Typically, effluent is treated and discharged from wastewater treatment plants, which may receive wastewater from such entities as households, factories, or commercial establishments.

[15] Lindane is also known as gamma-hexachlorocyclohexane (γ-HCH), gammaxene, and Gammallin. As a pharmaceutical, lindane is used in shampoos and creams to treat lice and scabies (mites).

[16] According to EPA officials, another pollutant for which EPA has issued national criteria – malathion – has pharmaceutical applications.

[17] NSTC was established by Executive Order in 1993. NSTC has multiple committees to address its broad responsibilities regarding the scientific and technical work of the executive branch. For example, CENRS advises and assists NSTC on federal research and development related to environment, natural resources, and sustainability. The Toxic and Risks Subcommittee is one of several subcommittees of CENRS.

[18] Congress established OSTP in 1976 with a broad mandate to advise the President and others within the Executive Office of the President on considerations of science and technology in federal policy, plans, and programs. OSTP is also charged with leading interagency efforts to develop and implement sound science and technology policies, among other things.

[19] Charter of the Working Group on Pharmaceuticals in the Environment, Toxics and Risk Subcommittee, Committee on Environment and Natural Resources, National Science and Technology Council (May 2006).

[20] The eight agencies were the U.S. Department of Agriculture, the Department of Commerce's National Oceanographic and Atmospheric Administration, EPA, the Department of Health and Human Services' Centers for Disease Control and Prevention, FDA, and National Institute for Environmental and Health Sciences; and the Department of the Interior's Fish and Wildlife Service and USGS.

[21] Focazio, M.J.; Kolpin, D.W.; Barnes, K.K.; Furlong, E.T.; Meyer, M.T.; Zaugg, S.D.; Barber, L.B.; Thurman, M.E., "A National Reconnaissance for Pharmaceuticals and Other Organic Wastewater Contaminants in the United States – II) Untreated Drinking Water Sources," Sci Total Environ, 402(2-3) (2008): 201-216.

[22] USGS selected these contaminants using the following criteria: known or suspected usage, toxicity, potential hormonal activity, persistence in the environment, and results from previous studies.

[23] Kolpin, D.W.; Furlong, E.T.; Meyer, M.T.; Thurman, E.M.; Zaugg, S.D.; Barber, L.B.; Buxton, H.T., "Pharmaceuticals, Hormones, and Other Organic Wastewater Contaminants in U.S. Streams, 1999-2000: A National Reconnaissance," Environ. Sci. Technol., 36(6) (2002): 1202-1211.

[24] Barnes, K.K.; Kolpin, D.W.; Furlong, E.T.; Zaugg, S.D.; Meyer, M.T.; Barber, L.B., "A National Reconnaissance of Pharmaceuticals and Other Organic Wastewater Contaminants in the United States – I) Groundwater," Sci Total Environ, 402(2-3) (2008): 192-200.

[25] Furlong, E.T.; Ferrer, I.; Glassmeyer, S.; Cahill, J.D.; Zaugg, S.D.; Werner, S.L.; Kinney, C.A.; Kolpin, D.W.; Kryak, D., "Distributions of organic wastewater contaminants between water and sediment in surface-water samples of the United States" (in Proceedings of the 3rd International Conference on Pharmaceuticals and Endocrine Disrupting Chemicals in Water, Minneapolis, Minn., National Ground Water Association, Mar. 19-21, 2003).

[26] Transport refers to the movement of a contaminant within the environment.

[27] Hinkle, S.R.; Weick, R.J.; Johnson, J.M.; Cahill, J.D.; Smith, S.G.; Rich, B.J., Organic Wastewater Compounds, Pharmaceuticals, and Coliphage in Groundwater Receiving Discharge from Onsite Wastewater Treatment Systems Near La Pine, Oregon: Occurrence and Implications for Transport, U.S. Geological Survey Scientific Investigations Report, 2005-5055 (2005): 98.

[28] NYC Department of Environmental Protection, Occurrence of Pharmaceutical and Personal Care Products (PPCP) in Source Water of the New York City Water Supply (New York, N.Y.: May 26, 2010).

[29] The National Water Research Institute (NWRI) is a nonprofit organization founded in 1991 by a group of California water agencies in partnership with the Joan Irvine Smith and Athalie R. Clarke Foundation to promote the protection, maintenance, and restoration of water supplies and to protect public health and improve the environment. NWRI's member agencies include Inland Empire Utilities Agency, Irvine Ranch Water District, Los Angeles Department of Water and Power, Orange County Sanitation District, Orange County Water District, and West Basin Municipal Water District.

[30] Y. Carrie Guox and Stuart W. Krasner, Metropolitan Water District of Southern California, La Verne, California; Steve Fitzsimmons, Greg Woodside, and Nira Yamachika, Orange County Water District Fountain Valley, California. Source, Fate, and Transport of Endocrine Disruptors, Pharmaceuticals, and Personal Care Products in Drinking Water Sources in California. A special report prepared at the request of the National Water Research Institute Fountain Valley, California, May 2010.

[31] According to its Web site, the American Water Works Association Research Foundation is a member-supported, international, nonprofit organization that sponsors research to enable water utilities, public health agencies, and other professionals to provide safe and affordable drinking water to consumers. Its stated mission is to advance the science of water to improve the quality of life.

[32] According to its Web site, the WateReuse Foundation is an educational, nonprofit public benefit corporation that serves as a centralized organization for the water and wastewater community to advance the science of water reuse, recycling, reclamation, and desalination.

[33] Benotti, M.J.; Trenholm, R.A.; Vanderford, B.J.; Holady, J.C.; Stanford, B.D.; and Snyder, S.A., "Pharmaceuticals and Endocrine Disrupting Compounds in U.S. Drinking Water," Environmental Science and Technology, 43(3) (2009): 597-603.

[34] Daughton C.G., "Pharmaceutical Ingredients in Drinking Water: Overview of Occurrence and Significance of Human Exposure," in ACS Symposium Series 1048, Contaminants of Emerging Concern in the Environment: Ecological and Human Health Considerations, ch. 2, Rolf Halden (ed.) (Washington, D.C.: American Chemical Society, November 2010).

[35] The study's analysis cautioned that the data cannot be considered statistically representative of any particular locale and that with very few exceptions; each of the 48 publications was very limited in scope. The analysis further cautions that the published studies employed various methods of analysis and quality control measures; and that no attempt was made to determine the veracity of the actual identification of targeted pharmaceuticals.

[36] Glassmeyer, S.T.; Furlong, E.T.; Kolpin, D.W.; Cahill, J.D.; Zaugg, S.D.; Werner, S.L.; Meyer, M.T.; Kryak, D.D., "Transport of Chemical and Microbial Compounds from Known Wastewater Discharges: Potential for Use as Indicators of Human Fecal Contamination," Environ Sci Technol, 39(14) (2005): 5157-5169.

[37] Antimicrobials are drugs used to treat infections by micro-organisms such as bacteria and viruses and include drugs such as synthetic and natural antibiotics.

[38] Campagnolo, E.R.; Johnson, K.R.; Karpati, A.; Rubin, C.S.; Kolpin, D.W.; Meyer, M.T.; Esteban, J.E.; Currier, R.W.; Smith, K.; Thu, K.M.; McGeehin, M., "Antimicrobial Residues in Animal Waste and Water Resources Proximal to Large-Scale Swine and Poultry Feeding Operations," Sci Total Environ, 299(1-3) (2002), 89-95.

[39] Phillips, P.J.; Smith, S.G.; Kolpin, D.W.; Zaugg, S.D.; Buxton, H.T.; Furlong, E.T.; Esposito, K.; Stinson, B., "Pharmaceutical Formulation Facilities as Sources of Opioids and Other Pharmaceuticals to Wastewater Treatment Plant Effluents," Environ. Sci. Technol., 44 (13) (2010): 4910-4916.

[40] While each of the studies cited in this report are limited in geographic scope and not generalizable to all public drinking water systems or sources of drinking water in the country, they still offer valuable insights. Please see the specific research citation for additional information.

[41] Statement by Alan Goldhammer, Ph.D., Deputy Vice President, Regulatory Affairs; Pharmaceutical Research and Manufacturers of America, before the Senate Environment and Public Works Committee, Subcommittee on Transportation Safety, Infrastructure Security, and Water Quality, "Pharmaceuticals in the Nation's Drinking Water: Assessing Potential Risks and Actions to Address the Issue" (Apr. 15, 2008).

[42] Schwab, B.W.; Hayes, E.P.; Fiori, J.M.; Mastrocco, F.J.; Roden, N.M.; Cragin, D.; Meyerhoff, R.D.; D'Aco, V.J.; Anderson, P.D., "Human Pharmaceuticals in U.S. Surface Water: A Human Health Risk Assessment," Reg. Toxicol. Pharmacol., 42(3) (2005): 296-312.

[43] Kolpin, D.W.; Furlong, E.T.; Meyer, M.T.; Thurman, E.M.; Zaugg, S.D.; Barber, L.B.; Buxton, H.T., "Pharmaceuticals, Hormones, and Other Organic Wastewater Contaminants in U.S. Streams, 1999-2000: A National Reconnaissance," Environ. Sci. Technol., 36(6) (2002): 1202-1211.

[44] Daughton C.G., "Pharmaceutical Ingredients in Drinking Water: Overview of Occurrence and Significance of Human Exposure," in Rolf Halden (ed.), ACS Symposium Series 1048, Contaminants of Emerging Concern in the Environment: Ecological and Human Health Considerations, ch. 2, Rolf Halden (ed.) (Washington, D.C.: American Chemical Society, November 2010).

[45] Daughton, C.G., "Pharmaceuticals as Environmental Pollutants: The Ramifications for Human Exposure," in the International Encyclopedia of Public Health, vol. 5, Kris Heggenhougen and Stella Quah (eds.) (Oxford: Academic Press, 2008): 66-102.

[46] Such subpopulations, which may be at greater risk for adverse health effects from exposure to drinking water contaminants, may include infants, children, individuals with kidney or liver diseases or weakened immune systems, and the elderly.

[47] Blazer, V.S.; Iwanowicz, L.R.; Iwanowicz, D.D.; Smith, D.R.; Young, J.A.; Hedrick, J.D.; Foster, S.W.; Reeser, S.J., 2007, "Intersex (Testicular Oocytes) in Smallmouth Bass from the Potomac River and Selected Nearby Drainages," Journal of Aquatic Animal Health, 19(4) (2007): 242-253.

[48] Kidd, K.A.; Blanchfield, P.J.; Mills, K.H.; Palace, V.P.; Evans, R.E.; Lazorchak, J.M.; Flick., R.W., "Collapse of a Fish Population after Exposure to a Synthetic Estrogen," Proceedings

of the National Academy of Sciences of the United States of America, 104(21) (2007): 8897-8901.

[49] Orlando, E. F.; Kolok, A. S.; Binzcik, G. A.; Gates, J. L.; Horton, M. K.; Lambright, C. S.; Gray, L. E.; Soto, A. M.; Guillette, L. J., "Endocrine-Disrupting Effects of Cattle Feedlot Effluent on an Aquatic Sentinel Species, The Fathead Minnow," Environ. Health Persp., 112(3) (2004): 353-358.

[50] Calafat, A.M.; Ye, X.; Wong, L.Y.; Reidy, J.A.; Needham, L.L. "Exposure of the U.S. Population to Bisphenol A and 4-tertiary-Octylphenol: 2003–2004." Environ. Health Perspect., 116(1) (January 2008): 39-44.

[51] Hayes, T.; Haston, K.; Tsui, M.; Hoang, A.; Haeffele, C.; Vonk, A., "Atrazine-Induced Hermaphroditism at 0.1 Ppb in American Leopard Frogs (Rana Pipiens): Laboratory and Field Evidence," Environ. Health Persp., 111 (2003): 568-575.

[52] According to studies published in 2008 and 2009, triclosan and triclocarban are also endocrine-disrupting chemicals. For example, see, Ahn, K.C.; Zhao, B.; Chen J.; Cherednichenko, G.; Sanmarti, E.; Denison, M.S.; Lasley, B.; Pessah, I.N.; Kultz, D.; Chang, D.P.Y.; Gee, S.J.; Hammock, B.D., "In Vitro Biologic Activities of the Antimicrobials Triclocarban, Its Analogs and Triclosan in Bioassay Screens: Receptor-Based Bioassay Screens," Environmental Health Perspectives, 116(9) (2008). Zorilla, L.M.; Gibson, E.K.; Jeffay, S.C.; Crofton, K.M.; Setzer, W.R.; Coper, R.L.; Stoker, T.E., "The Effects of Triclosan on Puberty and Thyroid Hormones in Male Wistar Rats," Toxicological Sciences, 107 (2009): 56-64.

[53] GAO, Antibiotic Resistance: Gaps in Information Will Remain Despite HHS Taking Steps to Improve Monitoring, GAO-11-406 (Washington, D.C.: June 1, 2011).

[54] For additional details on pollution prevention, source control, and other environmental stewardship measures see: Daughton, C.G, "Drugs and the Environment: Stewardship & Sustainability," National Exposure Research Laboratory, Environmental Sciences Division, U.S. EPA. Las Vegas, Nevada, report NERL-LV-ESD 10/081, EPA/600/R-106 (September 2010). Available at http://www.epa.gov/nerlesd1/bios/daughton/APM200-2010.pdf.

[55] The Product Steward Institute, Inc. is a nonprofit environmental organization whose members include 45 states, 70 local governments, and other stakeholders, and whose mission is to reduce the adverse health and environmental impacts of consumer products.

[56] The states are Arkansas, California, Colorado, Connecticut, Delaware, Florida, Idaho, Illinois, Indiana, Iowa, Kentucky, Maine, Massachusetts, Michigan, Missouri, New York, Oklahoma, Ohio, Oregon, Pennsylvania, Texas, Utah, Vermont, Washington, and Wisconsin.

[57] According to DEA regulations and the Controlled Substances Act, controlled substances include narcotics, stimulants, depressants, hallucinogens, anabolic steroids, and chemicals used in the illicit production of controlled substances. This list includes pharmaceuticals that are considered to have a high potential for abuse, such as opium, morphine, and methadone. DEA officials we spoke with stated that about 10 percent to 12 percent of all pharmaceuticals dispensed in the United States are controlled substances.

[58] Article 1, paragraph 87 of the European Union Directive 2004/27/EC (2004).

[59] GAO, Results-Oriented Government: Practices That Can Help Enhance and Sustain Collaboration among Federal Agencies, GAO-06-15 (Washington, D.C.: Oct. 21, 2005).

[60] Production volumes include information on the volume of pharmaceuticals manufactured in or imported into the United States in amounts equal to or greater than 10,000 pounds per year.

[61] In March 2011, EPA proposed the list of contaminants to be monitored under the third Unregulated Contaminant Monitoring Rule. EPA uses this rule to gather occurrence data on

unregulated contaminants in drinking water. The proposed list of contaminants for monitoring includes five of the pharmaceuticals on the 2009 Contaminant Candidate List. 76 Fed. Reg. 11713 (March 3, 2011)

[62] 73 Fed. Reg. 9628 (Feb. 21, 2008)

[63] Daughton C.G., "Pharmaceutical Ingredients in Drinking Water: Overview of Occurrence and Significance of Human Exposure," in ACS Symposium Series 1048, Contaminants of Emerging Concern in the Environment: Ecological and Human Health Considerations, ch. 2, Rolf Halden (ed.) (Washington, D.C.: American Chemical Society, November 2010).

[64] EPA Risk Assessment Forum, Guidance on Selecting Age Groups for Monitoring and Assessing Childhood Exposures to Environmental Contaminants (Washington, D.C., 2005).

[65] Kolpin, D.W.; Furlong, E.T.; Meyer, M.T.; Thurman, E.M.; Zaugg, S.D.; Barber, L.B.; Buxton, H.T., "Pharmaceuticals, Hormones, and Other Organic Wastewater Contaminants in U.S. Streams, 1999-2000: A National Reconnaissance," Environ. Sci. Technol., 36(6) (2002): 1202-1211.

[66] GAO-11-254.

[67] GAO, Homeland Defense: U.S. Northern Command Has Made Progress but Needs to Address Force Allocation, Readiness Tracking Gaps, and Other Issue, GAO-08-251 (Washington, D.C.: Apr.16, 2008); Live Animal Imports: Agencies Need Better Collaboration to Reduce the Risk of Animal-Related Diseases, GAO-11-9 (Washington, D.C.: Nov. 8, 2010).

[68] Charter of the Working Group on Pharmaceuticals in the Environment, Toxics and Risk Subcommittee, Committee on Environment and Natural Resources, National Science and Technology Council (May 26, 2006).

[69] According to the NSTC 2008 Handbook, NSTC documents must be cleared by the White House Co-Chair of the Committee or by the OSTP Director, or, in the event that neither is available, by the OSTP General Counsel.

[70] GAO-06-15.

End Note for Appendix II

[1] http://www.fda.gov/drugs/resourcesforyou/consumers/buyingusingmedicine safely/ensuringsafeuseofmedicine/safedisposalofmedicines/ucm186187. htm

End notes for Appendix III

[13] NSTC was established by Executive Order in 1993. NSTC has multiple committees to address its responsibilities regarding the scientific and technical work of the executive branch. For example, CENRS advises and assists NSTC on federal research and development related to environment, natural resources, and sustainability. The Toxic and Risks Subcommittee is one of several subcommittees of CENRS.

[14] Congress established OSTP in 1976 to advise the President and others within the Executive Office of the President on considerations of science and technology in federal policy, plans,

and programs. OSTP is also charged with leading interagency efforts to develop and implement sound science and technology policies, among other things.

[15] Charter of the Working Group on Pharmaceuticals in the Environment, Toxics and Risk Subcommittee, Committee on Environment and Natural Resources, National Science and Technology Council (May 26, 2006).

[16] The eight agencies were the U.S. Department of Agriculture, the Department of Commerce 's National Oceanographic and Atmospheric Administration, U.S. Environmental Protection Agency (EPA), the Department of Health and Human Services' Centers for Disease Control and Prevention, US. Food and Drug Administration (FDA), National Institute for Environmental and Health Sciences; and the Department of the Interior's Fish and Wildlife Service and U.S. Geological Survey (USGS).

[66] Charter of the Working Group on Pharmaceuticals in the Environment, Toxics and Risk Subcommittee, Committee on Environment and Natural Resources, National Science and Technology Council (May 26, 2006).

End Notes for Appendix IV

[13] NSTC was established by Executive Order in 1993. NSTC has multiple committees to address its responsibilities regarding the scientific and technical work of the executive branch. For example, CENRS advises and assists NSTC on federal research and development related to environment, natural resources, and sustainability. The Toxic and Risks Subcommittee is one of several subcommittees of CENRS.

[14] Congress established OSTP in 1976 to advise the President and others within the Executive Office of the President on considerations of science and technology in federal policy, plans, and programs. OSTP is also charged with leading interagency efforts to develop and implement sound science and technology policies, among other things.

[15] Charter of the Working Group on Pharmaceuticals in the Environment, Toxics and Risk Subcommittee, Committee on Environment and Natural Resources, National Science and Technology Council (May 26, 2006).

[16] The eight agencies were the U.S. Department of Agriculture, the Department of Commerce's National Oceanographic and Atmospheric Administration, U.S. Environmental Protection Agency (EPA). the Department of Health and Human Services' Centers for Disease Control and Prevention, U.S. Food and Drug Administration (FDA), National Institute for Environmental and Health Sciences; and the Department of the Interior's Fish and Wildlife Service and US. Geological Survey (USGS).

[66] Charter of the Working Group on Pharmaceuticals in the Environment, Toxics and Risk Subcommittee, Committee on Environment and Natural Resources, National Science and Technology Council (May 26. 2006).

In Pharmaceuticals in the U.S. Water Supply ISBN: 978-1-61942 -408-1
Editors: E. T. Blevins and S. E. Rush

Chapter 2

STATEMENT OF BARBARA BOXER: SUBCOMMITTEE ON TRANSPORTATION SAFETY, INFRASTRUCTURE SECURITY, AND WATER QUALITY HEARING ON, "PHARMACEUTICALS IN THE NATION'S WATER: ASSESSING POTENTIAL RISKS AND ACTIONS TO ADDRESS THE ISSUE"*

Tuesday, April 15, 2008

(Remarks as prepared for delivery)

We are here today to conduct much-needed oversight on the presence of pharmaceuticals in our nation's water supplies.

Clean, safe drinking water is essential to all of us. It is especially important for children.

The National Academy of Sciences has found that children drink more water and eat more food in proportion to their body weight compared to adults. In addition, children's rapidly developing bodies, including their hormone

* This is an edited, reformatted and augmented version of Statement given by Barbara Boxer, Hearing: Subcommittee on Transportation Safety, Infrastructure Security, and Water Quality hearing entitled, "Pharmaceuticals in the Nation's Water: Assessing Potential Risks and Actions to Address the Issue", on Tuesday, April 15, 2008.

systems that control their development, are especially vulnerable. Pregnant women also undergo a host of similarly delicate changes.

There are particular “windows of vulnerability” during our development when pregnant women and children may be especially susceptible to very low doses of some toxins.

Some pharmaceuticals now being found in our water may affect our hormone systems. Many pharmaceuticals are designed to affect our bodies at very low levels.

This all means that contaminants in water may have a more concentrated impact on pregnant women and children.

It is because of work by the United States Geological Survey, the Associated Press, and others, that we know that our water supplies can contain a mixture of pharmaceuticals.

Fish and wildlife that live in our waters are the familiar “canaries in a coal mine.” Scientific evidence is growing that small levels of contaminants, including pharmaceuticals, can damage reproduction and development in fish and wildlife. Science is telling us: be careful.

What are federal agencies doing to prevent potentially dangerous exposures to pharmaceuticals in our drinking water? The answer: very little.

EPA in particular has failed to adequately address this problem.

EPA has failed to require the needed testing to determine the effects of these chemicals at low levels. In 1996, Congress told EPA in the Safe Drinking Water Act and the Food Quality Protection Act to develop a program to identify and address chemicals that harm the natural balance of hormones in our body, called endocrine disrupting chemicals.

Yet, EPA is now nearly 6 years behind the schedule established in a court settlement to list the endocrine disrupting chemicals it will test. And EPA still has not even established all of the tests needed to detect these chemicals, much less evaluated the chemicals using those tests.

EPA now says it is not prepared to require drinking water systems to monitor for pharmaceuticals or to set standards for pharmaceuticals in tap water, because it doesn’t have data showing harmful effects from low levels of exposure. But this lack of data is in large part a result of EPA’s failure to ensure that companies that make these chemicals complete the testing needed to evaluate those effects.

Because of these problems, EPA has not set a drinking water standard for a single pharmaceutical. In fact, EPA hasn’t even proposed to set a single health standard for any pharmaceutical in drinking water.

The result of these failures is that millions of Americans turn on their kitchen taps and drink low levels of pharmaceuticals in their water every day.

The agency also should be doing much more to prevent these pharmaceuticals from getting into our water in the first place. For example, EPA should better address the disposal of pharmaceuticals, and the releases of these chemicals from factory farms, sewage treatment plants, and sewage sludge.

A White House working group that was supposed to address the pharmaceuticals in the environment and related antibiotic resistance issues has missed its deadline to issue recommendations. The [Bush] White House has insisted on keeping many records of this working group secret—which is unacceptable. All of the documents should be released to the public now.

The Associated Press story published in March documenting widespread drinking water contamination with pharmaceuticals highlights the importance and impact of the public right to know. When people are told about contaminants in their water, they rightfully demand action by their water suppliers and EPA.

Americans have a right to expect that their government is ensuring that they can turn on their taps and have water that is safe for their children and families to drink. And they have a right to know what is in their drinking water.

We must protect those who are most vulnerable—pregnant women, infants, and children—from this problem. That is our moral duty.

In Pharmaceuticals in the U.S. Water Supply ISBN: 978-1-61942 -408-1
Editors: E. T. Blevins and S. E. Rush

Chapter 3

STATEMENT OF DR. ROBERT M. HIRSCH U.S. GEOLOGICAL SURVEY BEFORE THE COMMITTEE ON ENVIRONMENT AND PUBLIC WORKS *

April 15, 2008

Mr. Chairman and Members of the Subcommittee, thank you for the opportunity to provide the views of the U.S. Geological Survey, Department of the Interior, on pharmaceuticals in the environment. The observed presence of pharmaceuticals in the environment has prompted public interest regarding potential adverse ecological effects and potential contamination of drinking water. The interest has already increased public awareness of the ways we handle and dispose of our medications and has resulted in interest by industries in waste treatment technologies and best management practices that are most effective at removing pharmaceuticals and other trace organic chemicals from surface and ground waters and solid and liquid wastes.

The USGS studies a wide range of chemicals of emerging environmental concern. Human and veterinary pharmaceuticals are only one class of chemicals in a new group of contaminants that enter the environment via human and animal wastes. These contaminants include many chemicals used

* This is an edited, reformatted and augmented version of Statement given by Dr. Robert M. Hirsch, Associate Director for Water U.S. Geological Survey U.S. Department of the Interior before the Committee on Environment and Public Works Subcommittee on Transportation Safety, Infrastructure Security and Water Quality, on April 15, 2008.

in our homes, businesses, and industries, including detergents, fragrances, fire retardants, disinfectants, plastics, and insect repellants.

Many of these chemicals are a new focus for environmental research, because they are used in relatively small quantities, and therefore, were not expected to be of significant environmental concern. In recent years, they have been detected increasingly in the environment at very low levels (less than one part per billion). Despite these extremely low levels, investigation is warranted to determine if there are any potential adverse environmental and human health effects. Research and monitoring by the USGS and others have demonstrated that

(1) the manner in which we handle and dispose of our wastes can concentrate these chemicals in some environmental settings to levels that may be an ecological health concern, and
(2) pharmaceuticals have been entering the environment for as long as we have used them.

In 1998, the USGS broadened its water-quality science programs by initiating research on pharmaceuticals and other human- and animal-waste related chemicals. We were spurred by the findings of European colleagues, who, looking for a pesticide, detected a heart medication in the North Sea (Buser et al., 1998). The realization that chemical-use and waste-handling practices had resulted in detectable concentrations of a drug in such a large water body suggested the need for further research. By 2002, a USGS study (Kolpin et al., 2002; Barnes et al., 2002; and Buxton and Kolpin, 2002) had documented the presence of pharmaceuticals and other waste-associated chemicals in the Nation's streams, and largely defined this issue in the United States.

Since 2002, the USGS has published more than 160 reports that:

- Document the occurrence, concentration, and mixtures of these chemicals in various environmental compartments, including stream water, well water, stream sediment, and soil amended with manure and biosolids (the solid byproduct of wastewater treatment plants);
- Demonstrate the comparative contributions from various sources, including wastewater treatment plants, livestock production and animal feedlot wastes, aquaculture, onsite septic systems, combined sewer overflows, and other industrial discharges; and

- Demonstrate that some of these chemicals are assimilated by organisms (Kinney et al., 2008) or cause adverse ecological health effects (Vajda et al., 2008).

The bibliography of USGS reports that support these findings on pharmaceuticals and other emerging contaminants in the environment is available on the Internet at: *http://toxics.usgs.gov/bib/bib-Emerging.html.*

A recent example of USGS research is described in a series of reports on the levels and mixtures of human- and animal-waste related chemicals that are found in wastewaters, biosolids, and manures, and the soils to which they are applied for fertilization, as well as the earthworms found in those soils (Kinney et al., 2006a, 2006b, and 2008).

USGS investigations at a conventional drinking water treatment facility in New Jersey described the changes in pharmaceutical concentrations from the source water through multiple stages of the treatment process (Stackelberg et al., 2004, 2007). Additional investigations like this one will inform decisions on improving existing and developing new treatment works that are more efficient at removing these compounds from source waters (the sources of drinking water).

Two USGS papers, anticipated to be published in the coming weeks in the peer-reviewed journal *Science of the Total Environment,* summarize the occurrence of these chemicals in ground water (Barnes et al., in press) and in raw (untreated) sources (streams and wells) of drinking water (Focazio et al., in press). The paper surveying source waters includes results from 74 sites near drinking water intakes in 25 states and Puerto Rico. All data from these reports will be made available to the public in accompanying data reports available on the USGS web site.

The ecological effects of some pharmaceuticals found in the environment have been documented in scientific literature For example, it was not a surprise when antibiotics, which are designed to act as antibacterials, were found to have adverse effects on soil microorganisms at environmentally relevant concentrations (Thiele-Bruhn and Beck, 2005). Testing also has found that some pharmaceuticals do not cause adverse effects on some organisms tested. In one study, three freshwater invertebrates were exposed to an anticonvulsant drug commonly found in the environment. Only one of the 3 species demonstrated an adverse effect (Oetken et al., 2005). Furthermore, evidence suggests that chemical mixtures can act collectively to cause adverse effects, even when each component is below its individual effect level (Brian et al., 2007). These are examples of an increasing body of scientific knowledge on potential adverse health effects. Significant uncertainty remains regarding the effects of long-term exposure to levels found in environmental settings.

Endocrine disruption is one adverse health effect of concern because it may occur as a result of exposure to very low levels of hormonally active chemicals. One form of endocrine disruption observed in environmental settings affects fish reproductive systems, where fish have been found to be "feminized" by exposure to a range of chemicals that act similar to the natural hormone estrogen. Some ways in which feminization is observed in fish include: (1) elevation in the percent of fish populations that are female, (2) changes in behavioral characteristics, such as nesting behavior, or (3) the presence of male fish with female characteristics, such as the presence of female egg cells in testes or of a female egg protein in their blood. A recent study (Kidd et al., 2007) demonstrated that the addition of ethinylestradiol (one of the active ingredients in birth control pills) at observed environmental concentrations to an experimental lake in Canada caused feminization and near extinction of fathead minnows in the lake.

A wide range of hormonally active chemicals can contribute to endocrine disruption, including actual biogenic hormones, synthetic hormones (pharmaceuticals, such as ethinylestradiol), and other chemicals that mimic or block hormone function (including certain pesticides, detergents, metals, and other industrial chemicals). These chemicals have been found together in waters affected by human and animal wastes and can occur together in various environmental settings. This reinforces why these chemicals must be studied together and not as separate classes of contaminants.

The effects of long-term exposure to the low levels of pharmaceuticals found in the environment on human health are not understood and warrant continued study. The USGS has collected information on the occurrence, concentrations, and mixtures of pharmaceuticals and other waste-related chemicals in source waters used for drinking water and, to a much more limited extent, in finished (treated) drinking water. Information provided to the Food and Drug Administration (FDA) as part of the drug approval process can be useful in identifying more potent pharmaceuticals as the target of future research on possible effects of pharmaceuticals in the environment. However, whether or not there are adverse human health effects from cumulative lifetime exposures to the low concentrations and complex mixtures of pharmaceuticals found in the environment remains a research priority, particularly the effects on sensitive subpopulations such as children, women of child-bearing years, the elderly, and people with suppressed immune systems.

The USGS has developed the capability to analyze for approximately 70 pharmaceuticals in environmental samples. We have collected and analyzed samples from approximately 1,500 sites across the Nation. About a quarter of these sites were sampled in nationally-designed targeted surveys implemented by the USGS to assess the occurrence of pharmaceuticals across a wide range of environmental settings. The majority of the 1,500 sites were sampled as parts of studies conducted by the USGS in cooperation with State and local governments. These cooperative studies were designed to provide information for local resource managers on conditions in their area, and the findings are available to the public.

California provides an example of an extensive State program. The USGS California Water Science Center, in collaboration with the California Water Boards, has designed and implemented the Groundwater Ambient Monitoring and Assessment (GAMA) Project to assess the quality of California's ground water. The Priority Basin Program, a part of GAMA, will sample approximately 2,500 wells in about 120 ground-water basins in California over an 8-year period (2004-2012). From May 2004 through December 2007, we have sampled about 1,400 wells for a very large suite of constituents, both natural and human-generated, including pharmaceuticals. About 1,000 wells are being evaluated for the presence of pharmaceuticals.

The USGS is continuing to conduct research on pharmaceuticals in the environment. Our research priorities will continue to include assessing:

- Chemical loads of various sources including wastewater treatment plants, Animal Feeding Operations, landfills, and other industrial facilities,
- Ecological effects, including fish endocrine disruption in streams enriched with wastewaters and antimicrobial resistance in settings where antibiotics are released to the environment,
- The occurrence of pharmaceuticals in waters that are the source of drinking water and, to a limited degree, in treated drinking water, and
- The comparative performance of varying water and waste treatment processes to remove pharmaceuticals and other chemicals,

The USGS conducts research on pharmaceuticals in the environment with a number of partner Federal agencies, including the U.S. Environmental Protection Agency (USEPA), Centers for Disease Control and Prevention, Fish and Wildlife Service, and National Oceanic and Atmospheric

Administration. The USGS, USEPA, and FDA co-chair the Federal Interagency Work Group on Pharmaceuticals in the Environment, under the auspices of the Committee on Environment and Natural Resources of the National Science and Technology Council. The Work Group has further increased coordination of Federal research. The USGS is working with our partner Federal agencies, including the FDA, to prioritize pharmaceuticals for environmental study.

Thousands of pharmaceuticals are used for human therapies and veterinary purposes. The USGS is focusing environmental research on chemicals that are more likely to be of environmental concern, to increase the efficiency of research within the existing limited resources available. Similarly, investigations of adverse health effects must consider the actual levels and mixtures of chemicals that organisms are exposed to in the environment. Results of USGS studies of environmental occurrence are used by many scientists to guide both human and ecological health-effects studies to assure that actual environmental conditions are being tested.

We welcome the opportunity to provide any further information or assistance to the Subcommittee. Thank you, Mr. Chairman, for the opportunity to present this testimony, and I will be pleased to answer questions you and other Members might have.

References

Barnes KK, Kolpin DW, Meyer MT, Thurman EM, Furlong ET, Zaugg SD, and Barber, LB, 2002, Water-quality data for pharmaceuticals, hormones, and other organic wastewater contaminants in U.S. streams, 1999-2000: U.S. Geological Survey Open-File Report 02-94. http://toxics.usgs.gov/pubs/OFR-02-94/

Barnes KK, Kolpin DW, Furlong ET, Zaugg SD, Meyer MT, Barber LB, (in press), A national reconnaissance for pharmaceuticals and other organic wastewater contaminants in the United States: I) Ground water. *Science of the Total Environment.*

Brian JV, Harris CA, Scholze M, Kortenkamp A, Booy P, Lamoree M, Pojana G, Jonkers N, Marcomini A, and Sumpter JP, 2007, Evidence of estrogenic mixture effects on the reproductive performance of fish. *Environmental Science and Technology, 41*(1), 337- 344.

Buser HR, Muller MD, Theobald N, 1998, Occurrence of the pharmaceutical drug clofibric acid and the herbicide mecoprop in various Swiss lakes and in the North Sea. *Environmental Science & Technology*, *32*, 188-192.

Buxton HT, and Kolpin DW, 2002, Pharmaceuticals, hormones, and other organic wastewater contaminants in U.S. streams: U.S. Geological Survey Fact Sheet FS-027-02, 2 p. http://toxics.usgs.gov/pubs/FS-027-02/

Focazio MJ, Kolpin DW, Barnes KK, Furlong ET, Meyer, MT, Zaugg, SD, Barber, LB, Thurman, EM, (in press). A national reconnaissance for pharmaceuticals and other organic wastewater contaminants in the United States: II) Untreated drinking water sources. *Science of the Total Environment.*

Kidd KA, Blanchfield PJ, Mill KH, Palace VP, Evans RE, Lazorchak JM, Flick RW, 2007, Collapse of a fish population after exposure to a synthetic estrogen. *Proceedings of the National Academy of Sciences*, *104*, 8897-8901.

Kinney CA, Furlong ET, Werner SL, Cahill JD, 2006a, Presence and distribution of wastewater-derived pharmaceuticals in soil irrigated with reclaimed water. *Environmental Toxicology and Chemistry*, *25*, 317-326. http://pubs.acs.org/cgibin/abstract.cgi/esthag/2008/ 42/i06/abs/ es702304c. html

Kinney CA, Furlong ET, Zaugg SD, Burkhardt MR, Werner SL, Cahill JD, Jorgensen GR, 2006b, Survey of organic wastewater contaminants in biosolids destined for land application. *Environmental Science & Technology*, *40*, 7207-7215. http://www. setacjournals.org/perlserv /?request=get-abstract&doi=10.1897%2F05-187R.1

Kinney CA, Furlong ET, Kolpin, DW, Burkhardt MR, Zaugg SD, Werner SL, Bossio, JP, and Benotti, MJ, 2008, Bioaccumulation of pharmaceuticals and other anthropogenic waste indicators in earthworms from agricultural soil amended with biosolid or swine manure. *Environmental Science and Technology*, *42* (6), 1863–1870. http://pubs.acs.org/cgi-bin/abstract.cgi/ esthag/2006/40/i23/abs/es0603406.html

Kolpin DW, Furlong ET, Meyer MT, Thurman EM, Zaugg SD, Barber LB, Buxton HT, 2002, Pharmaceuticals, hormones, and other organic wastewater contaminants in US streams, 1999-2000: A national reconnaissance. *Environmental Science & Technology*, *36*, 1202-1211. http://pubs.acs.org/hotartcl/est/es011055j rev.html

Oetken M, Nentwig G, Loffler D, Ternes T, Oehlmann J, 2005, Effects of pharmaceuticals on aquatic invertebrates. Part I. The antiepileptic drug carbamazepine. *Archives of Environmental Contamination and Toxicology 49*, 353-361.

Stackelberg PE, Gibs J, Furlong ET, Meyer MT, Zaugg SD, and Lippincott RL, 2007, Efficiency of conventional drinking-water-treatment processes in removal of pharmaceuticals and other organic compounds. *Science of the Total Environment*, v. 377, no. 2-3, p. 255-272.

http://www.sciencedirect.com/science? ob=ArticleURL& udi=B6V78-4NB2SFN2&_user=696292&_rdoc=1&_fmt=&_orig=search&_sort =d&view=c& acct=C0000388 19&_version=1&_urlVersion=0&_ userid=696292&md5= 782f1fc5e 594b3d0e696dfd 62b bf79fd

Stackelberg PE, Furlong ET, Meyer MT, Zaugg SD, Henderson AK, and Reissman DB, 2004, Persistence of pharmaceutical compounds and other organic wastewater contaminants in a conventional drinking-water-treatment plant. *Science of the Total Environment*, v. 329, no. 1-3, p. 99-113.

http://www.sciencedirect.com/science?_ob=ArticleURL&_udi=B6V78-4CC30H2- K&_user=696292&_rdoc=1&_fmt=&_orig=search&_sort=d& view=c&_acct=C0000388 19&_version=1&_urlVersion=0&_userid =696 292&md5=efb0c0e5ad401e84e 2a4511 c9557f4cc

Thiele-Bruhn S, and Beck I, 2005, Effects of sulfonamide and tetracycline antibiotics on soil microbial activity and microbial biomass. *Chemosphere*, vol. 59, pp. 4457-465.

Vajda, AM, Barber, LB, Gray, JL, Lopez, EM, Woodling, JD, and Norris, DO, 2008, Reproductive disruption in fish downstream of an estrogenic wastewater effluent: *Environmental Science & Technology* (published on-line March 25, 2008).

In Pharmaceuticals in the U.S. Water Supply ISBN: 978-1-61942 -408-1
Editors: E. T. Blevins and S. E. Rush

Chapter 4

TESTIMONY OF BENJAMIN H. GRUMBLES, ENVIRONMENTAL PROTECTION AGENCY, BEFORE THE TRANSPORTATION SAFETY, INFRASTRUCTURE SECURITY AND WATER QUALITY SUBCOMMITTEE OF THE ENVIRONMENT AND PUBLIC WORKS COMMITTEE UNITED STATES SENATE*

April 15, 2008

Good afternoon Chairman Lautenberg and members of the Committee. I am Benjamin H. Grumbles, Assistant Administrator for Water at the United States Environmental Protection Agency (EPA). I appreciate the opportunity to describe EPA's actions to evaluate the potential risks to human health and aquatic life posed by trace amounts of pharmaceuticals in water, and to identify measures to minimize their occurrence in water. The Agency is committed to undertaking a scientific approach in evaluating the risks associated with contaminants in our environment so that we can take the necessary action to ensure clean and safe water.

* This is an edited, reformatted and augmented version of testimony given by Benjamin H. Grumbles Assistant Administrator for Water Environmental Protection Agency before the Transportation Safety, Infrastructure Security and Water QualitySubcommittee of the Environment and Public Works Committee United States Senate, on April 15, 2008.

EPA is concerned about the detection of a number of pharmaceuticals and personal care products in our water. EPA has been actively working with federal agencies and state and local government partners to better understand the implications of emerging contaminants such as pharmaceuticals, endocrine disrupting chemicals, and personal care products detected in drinking water, wastewater, surface water and ground water. We continue to evaluate their routes of exposure, levels of exposure, and potential effects on public health and aquatic life.

Over the last few years, EPA has increased its work in a number of areas to better understand pharmaceuticals and personal care products. We are focused on learning more about the occurrence of pharmaceuticals and personal care products in water. In addition, we are working to better understand what treatment technologies may remove them from wastewater and drinking water. We are developing analytical methods to improve detection capabilities. We are conducting national studies and surveys to help direct our course of action. We are also partnering with government agencies, stakeholders, and the private sector, and increasing public awareness about product stewardship and pollution prevention.

U.S. Drinking Water is among the Safest in the World

The U.S. has one of the safest drinking water supplies in the world. EPA is committed to keeping our water clean and healthy for the future.

While there is much information about the health effects of pharmaceutical products at the therapeutic doses provided in medication, there is still uncertainty about their potential effects on public health and aquatic life at the extremely low levels observed in drinking water and surface water. Let me give you an example.

Although pharmaceuticals have been detected in a number of U.S. waters (as well as waters worldwide), the amounts at which they are detected -- in the parts per billion and parts per trillion range --are extremely low. For example, in looking at a number of studies, the maximum reported drinking water level for caffeine was 0.12 parts per billion. At that concentration, a person would have to drink almost 222,000 gallons of water before coming close to ingesting the amount of caffeine that one would get from a six ounce cup of coffee.

We know that collaborating with our partners will be critical so that we can all make the best use of our existing resources and maintain consumer confidence in our drinking water. Keeping public drinking water safe and ensuring that consumers have high confidence in our nation's drinking water is an absolute tenet of EPA's approach to environmental and public health protection. It is also testimony to the daily efforts of thousands of people from the federal officials who develop standards, to the state managers that carry out programs, and the local officials who provide drinking water and treat wastewater.

EPA's Four-Pronged Approach to Emerging Contaminants

EPA is responding to emerging contaminants, with a four-pronged approach aimed at improving science, communicating risks, identifying partnership and stewardship opportunities, and preparing to take regulatory action when appropriate.

Strengthening the Science

Sound science and reliable information must be the foundation for any Agency decision. There is critical work that needs to be done in the area of research and assessment before we can make any decisions as to whether regulatory actions are needed. EPA and other federal agencies are working on projects to evaluate exposure and potential health effects on humans and aquatic life. This is a key, because while we know that pharmaceuticals have health effects at the therapeutic dose, we are less certain about the health effects associated with long-term exposure to much lower concentrations. Effects may be more likely in aquatic life because they are continually exposed.

Several EPA offices, including my Office and the Office of Research and Development, are working together to better understand potential issues related to exposure pathways and health effects of emerging contaminants including pharmaceuticals. I will describe some of the efforts we are undertaking in my office and have provided a brief summary of efforts in the Office of Research and Development as an attachment to this testimony.

EPA's Office of Research and Development is engaged in a broad portfolio of research efforts, much of which is focused on answering key questions associated with exposure pathways and health effects. EPA is also coordinating research efforts with other federal agencies as part of the Pharmaceuticals in the Environment (PiE) workgroup, under the auspices of the White House's National Science and Technology Council Committee on Environment and Natural Resources Toxics and Risk Subcommittee. The workgroup is co-chaired by EPA, the U.S. Geological Survey, and the Food and Drug Administration.

Other key research areas are associated with assessing the sources of pharmaceuticals. Important to this effort is having analytical methods available to reliably detect these contaminants in water. It is critical to ensure sound collection and analysis of samples because there is a great potential for cross-contamination and thus false identification and incorrect concentrations at the low levels detected in most samples. It is also important to note that the equipment needed to test for pharmaceuticals is highly technical and that analyses are very expensive relative to other contaminants. In December 2007, the Agency released newly developed, cutting edge methods for the analysis of approximately 100 pharmaceuticals, personal care products, steroids, and hormones in raw sewage, treated wastewater, and biosolids which are some of the most complex samples to test. The new methods, which are continuing to be refined, use high performance liquid chromatography combined with tandem mass spectrometry. The availability of the methods responds to requests for guidance in this area and the need to support studies being conducted by the Agency. We will continue to consider and evaluate additional methods.

EPA, other federal agencies, and academic and private sector researchers are studying the occurrence of pharmaceuticals in wastewater, surface water, ground water and drinking water. We are also evaluating the occurrence of contaminants in fish and other aquatic life. In particular, EPA's Office of Water is conducting:

- a study at nine wastewater treatment facilities to better understand what is going into the plant for treatment and what is coming out in the discharge and in biosolids (sludge). We expect to complete this study by December 2009.
- a Fish Tissue Pilot Study to investigate whether pharmaceuticals and other personal care products may occur in fish from five effluent-

dominated streams across the US. Results are undergoing quality assurance review and are expected later this year.

- a survey of biosolids from 74 randomly selected wastewater treatment plants to determine whether contaminants occur in biosolids, and if so at what concentration.

My office has also funded external research projects to assess occurrence. Researchers at the University of Florida are studying the fate and transport of emerging contaminants like triclocarban (an antiseptic widely used in soaps and other products) in biosolids. Researchers at Duke University are studying the presence, fate and treatability of steroid and hormone contaminants in wastewater and biosolids. They are conducting a year-long field study in four municipal wastewater treatment plants and are evaluating the overall potency of the wastewater through laboratory tests.

Additionally, EPA's Office of Research and Development's research agenda includes work to better understand pharmaceuticals in water through several important programs. A brief summary of those activities is attached to this testimony.

We will continue to work within EPA and with our federal, state, and local partners to identify and analyze available information and potential studies to assess the occurrence of pharmaceuticals in our drinking and surface water, along with any potential risks to human health and aquatic organisms.

Another important research area is treatment and removal of pharmaceuticals from wastewater and drinking water. While EPA is active in this area, research foundations representing drinking water and wastewater utilities are key players in determining the removal efficiency of different types of treatment. Research is finding that more sophisticated higher-level treatment strategies are sometimes better at removing certain types of pharmaceuticals. However, it is important to ensure that utilities continue focusing their efforts at removing those contaminants with known risks (such as pathogens), especially when monitoring and/or treatment efforts aimed at pharmaceuticals at low levels could carry significant cost with unknown risk reduction.

Improving Public Understanding and Risk Communication

One of the most difficult tasks faced by public health and environmental officials is how to communicate risks in the face of uncertainty. At this time we simply do not know whether there is a human health risk of concern from the levels that have been reported in water. Some have argued that it does not make sense to monitor for pharmaceuticals in water if there is limited information about the health effects at the concentrations that could be detected. We disagree. Information about occurrence and health effects is complementary and should be developed in tandem. It would be unfortunate indeed if proactive utilities referenced in media articles made decisions to reduce their research in response to media attention. Furthermore, useful information should be shared with the public in a timely way as it is generated. It is important to communicate with the public so that they can help shape effective public policy in this area and make informed choices. We will continue to work with all of our stakeholders to do the best job possible in communicating available data on both occurrence and health effects for all emerging contaminants, including pharmaceuticals, and any associated uncertainty with that data.

Building Partnerships for Stewardship

One thing we know for certain is that this issue cannot be addressed by EPA alone. Other federal, state, and local agencies and industry will also need to play a role in assessing the occurrence and effects of pharmaceuticals, analyzing their risk, and in determining actions to reduce their concentration in the environment. EPA is already coordinating research efforts with seven other federal agencies as part of the PiE workgroup which is examining current efforts with regard to human and veterinary pharmaceuticals in the environment, , with the goal of avoiding duplication of effort, leveraging existing resources, and better prioritizing Federal efforts.

We also worked with the White House Office of National Drug Control Policy to develop joint guidelines that recommend appropriate disposal methods for unused medication. EPA, other agencies, and the pharmaceutical industry are involved in a number of stewardship activities across the country to communicate the guidelines to the public. While most pharmaceuticals are entering water through natural biological functions, it is also important that the public understand that the toilet is not a trash can for unused medications.

EPA has also been working to develop and promote good stewardship efforts such as take-back programs that would allow consumers to properly dispose unwanted or unused pharmaceuticals. EPA recognizes that any such programs must be consistent with the Controlled Substances Act and regulations with respect to managing medications that are also classified as controlled substances. Toward this end, the Agency will be working with the Drug Enforcement Administration (DEA) to ensure that pilot take-back programs supported by EPA are conducted in a manner that is safe and in compliance with federal and state laws and regulations.

One such program that EPA recently funded involves the Area Resources for Community and Human Services in St. Louis. Last year, EPA's Office of Children's Health Protection and Environmental Education provided grants to this community partnership, which is working on an efficient regional model to responsibly dispose of unwanted non-controlled medications through a regional grocery store chain as the collection point. We are working on the evaluation and development of this pilot return take back programs to allow consumers to return and dispose of medications in a safe manner that complies with federal and state law and are working closely with the Drug Enforcement Administration to ensure applicability to all laws.

Additionally, EPA provided a grant to Villanova University to identify ways to better manage how prescription and non-prescription pharmaceuticals are discarded from university dormitories. The results of Villanova's work can be useful for other universities that are voluntarily taking steps to reduce pollution on their campuses as their commitment to improving the environment.

USING REGULATORY TOOLS

We recognize stewardship activities alone are not always sufficient to manage issues associated with emerging contaminants in water. We are also gathering information that will help us assess whether direct regulatory action is warranted. For example, under the Clean Water Act, EPA establishes technology-based national regulations, termed "effluent guidelines," to reduce pollutant discharges from categories of industrial facilities to waters of the United States. As part of the effluent guidelines planning process, the Agency is reviewing the pharmaceutical disposal practices of hospitals and long-term health care facilities. We expect to issue a report on our findings in 2009. We view this as an important opportunity to increase product stewardship and proper waste disposal.

Under the Safe Drinking Water Act, the Agency carries out a program to assess contaminants for potential drinking water regulation. On February 21, 2008, the Agency released the draft Contaminant Candidate List (CCL 3) for public review and comment. As part of the process to develop the list, the Agency evaluated pharmaceuticals and personal care products to identify those that have the potential to occur in drinking water provided by public water systems. EPA considered 287 chemicals identified as pharmaceuticals and personal care products; however, only one was included on the Draft CCL 3 because most occurred at levels far below those currently associated with any adverse health effects, based on the best available human health effects data. The Agency is seeking additional data and information on the concentrations of pharmaceuticals in treated or ambient water and adverse health effects that may be posed by their presence. We are also interested in receiving feedback on how we considered pharmaceuticals within the CCL 3 process.

The CCL 3 process evaluated many types of contaminants that can occur in drinking water – including microbial pathogens, pesticides, and chemicals used in industrial practices and consumer products. The 104 contaminants on the Draft CCL 3 include 93 chemicals and 11 microbiological contaminants. In the absence of reliable data indicating potential risks associated with pharmaceuticals in water at the very low levels at which they have been detected, it would be inappropriate to require monitoring and/or treatment that could carry significant cost, with no evidence of significant risk reduction based on currently available data. The Agency needs to instead focus its regulatory resources on contaminants with known significant risks in order to maximize public health protection.

Looking to the Future

EPA will continue to work on the important issue of pharmaceuticals in water. We regard this as an issue that warrants additional, on-going scrutiny.

I have sent letters to the directors of state environmental and health agencies to inform them about some of our efforts and to request their assistance. In conjunction with the CCL 3 comment period, I have asked them whether they are currently implementing, or planning to implement, a program to monitor for pharmaceuticals and personal care products in wastewater, surface water, ground water, or tap water. This type of information can be very useful to EPA as we carry out our contaminant candidate listing process to identify potential contaminants for unregulated contaminant monitoring and/or drinking water regulation, revise effluent guidelines, and determine which contaminants are the highest priorities for development of new or revised water quality criteria.

As part of EPA's broader strategy to strengthen and expand technical partnerships and information sharing at all levels, I have also asked the states to share information with us with respect to stewardship activities they may be undertaking to manage the presence of pharmaceuticals and personal care products in water within their state. We will compile the information we receive with the goal of sharing best practices and encouraging broad adoption of effective programs across the country.

Some of the additional activities we are considering include:

- expanding our current fish tissue pilot study to ensure a statistically and nationally representative picture of the presence of pharmaceuticals and personal care products in aquatic life;
- improving our public access web site to provide information on the work that we are doing and to communicate our understanding of risks to human and aquatic health;
- working within EPA and across federal agencies to better understand stewardship efforts they are undertaking to both improve coordination and share information; and,
- working to obtain toxicology data from available sources (including other federal agencies) and exploring ways to improve understanding of health effects posed by exposure to low levels of pharmaceuticals in water.

I will also be meeting with a number of key stakeholder groups over the next several weeks – including those representing state programs, the water and wastewater industry, research associations, and the environmental community. It is essential for us to share information and understand specific concerns so that we can work together as effectively as possible in communicating with the public.

Finally, while I have primarily discussed Office of Water activities here today, other offices within EPA and across the EPA Regions have also been carrying out efforts associated with pharmaceuticals and other emerging contaminants. They are reviewing their efforts to date and assessing what additional activities they may undertake in the future.

CONCLUSION

The U.S. continues to have one of the safest drinking water supplies in the world due to our collective long-standing commitment to keep our water clean and healthy. We know that good science and information must continue to drive our decisions. We will continue to evaluate health effects, occurrence, and risk reduction strategies so that we can make sound decisions to protect public health and aquatic life. We recognize the public must be a full partner in our efforts to collect and evaluate data so they can help shape effective public policy in this area and make informed choices. By engaging the full range of public and private partners and by using appropriate regulatory tools, we will continue to ensure the safety of the nation's water.

Thank you Chairman Lautenberg and members of the Committee for this opportunity to describe EPA's important work on pharmaceuticals and other emerging contaminants. I would be happy to answer any questions you may have.

In Pharmaceuticals in the U.S. Water Supply ISBN: 978-1-61942 -408-1
Editors: E. T. Blevins and S. E. Rush

Chapter 5

TESTIMONY OF JENNIFER SASS, NATURAL RESOURCES DEFENSE COUNCIL BEFORE THE US SENATE COMMITTEE ON ENVIRONMENT AND PUBLIC WORKS SUBCOMMITTEE ON TRANSPORTATION SAFETY, INFRASTRUCTURE SECURITY, AND WATER QUALITY AT HEARING ON: PHARMACEUTICALS IN THE NATION'S WATER: ASSESSING POTENTIAL RISKS AND ACTIONS TO ADDRESS THE ISSUE*

April 15, 2008

Good morning and thank you for this opportunity to testify on the health concerns and policy proposals addressing pharmaceuticals and personal care products in the nation's waterways and drinking water sources. I am Jennifer

---

* This is an edited, reformatted and augmented version of testimony given by Jennifer Sass, Ph. D Senior Scientist Natural Resources Defense Council on Behalf of: Natural Resources Defense Council before the Us Senate Committee on Environment and Public Works Subcommittee on Transportation Safety, Infrastructure Security, and Water Quality at Hearing Entitled: Pharmaceuticals in the Nation's Water: Assessing Potential Risks and Actions to Address the Issue, on April 15. 2008.

Sass, PhD., a Senior Scientist in the Health Program at the Natural Resources Defense Council (NRDC). I have a Ph.D. in molecular and developmental biology, and a post-doctoral certificate in environmental toxicology. I have worked at NRDC on environmental health issues for over 7 years. NRDC is a not-for-profit environmental advocacy organization with over 1 million members and activists whose mission is to safeguard the Earth: its people, its plants and animals and the natural systems on which all life depends.

NRDC's Health program focuses on toxic chemical pollutants in air, water, food, and shelter. Over the years, we have focused our particular attention on the "biggest pollutants" in these media, the ones disproportionately responsible for the biggest threats to human health. This has led to successful efforts to substantially reduce diesel air emissions from trucks and buses, for example, and to take a number of dangerous and outdated pesticides off the market. There are more than 70,000 chemicals in commerce, but some are much more toxic than others, and we can make great progress in environmental health protection if we focus on the chemicals pollutants that pose the greatest threat to human and ecological health.

We are very pleased to testify today on the health risks posed to humans and wildlife by pharmaceuticals and personal care products that contaminate our waterways.

Summary of Testimony

Compounds such as nicotine, caffeine, and aspirin that are designed to influence our body's normal chemistry have been identified as environmental contaminants since the 1980s, moving from sewage and human waste into waterways. As our use of pharmaceuticals increases, it is logical to expect them to turn up in our environment. Although the levels reported to contaminate our waterways are much lower than therapeutic doses, it would be naïve to think of this as 'safe', knowing that the agents are chemically reactive in our bodies, and that we are exposed daily over a life-time to multiple compounds in unknown combinations.

When a medical professional prescribes a drug, they are considering the patient's health status, age, gender, nutritional status, and any other drugs that may cross-react. For example, a woman who was at risk for breast cancer would not want to be exposed to high levels of estrogenic compounds. A pregnant woman would never knowingly expose her fetus to chemicals that cause birth defects such as antiepileptic drugs. A doctor would never knowingly prescribe toxic chemotherapy agents to a healthy person. Yet, all these things and more are in our Nation's drinking water. The Associated Press reported that pharmaceutical residues were detected in the drinking water of 24 major metropolitan areas across the country serving 41 million people. Detected drugs included antibiotics, anti-convulsants, and mood stabilizer drugs. These results were supported by findings of the U.S. Geological Survey that found organic wastewater contaminants and pharmaceuticals in 80% of sampled streams- including antibiotics, hypertensive and cholesterol-lowering drugs, antidepressants, analgesics, steroids, caffeine, and reproductive hormones.

Pharmaceuticals and personal care products (PPCPs) may end up in the environment through waste from human or animal excretion, improper disposal such as flushing down a toilet, runoff from animal feeding operations, or leaching from municipal landfills. However they get there, they are contaminating our waterways and tap water systems.

Large animal feeding operations generate a large amount of antibiotic-contaminated waste that contaminates waterways and contributes to antibiotic resistant pathogens. Because many of the same antibiotics are used in both human and veterinary medicine, almost every bacteria that can cause infections in humans has developed resistance to at least one antibiotic, and some are resistant to multiple antibiotics.

Pharmaceuticals that mimic estrogen are excreted as waste by-products from the use of birth-control pills, menopause treatments, and cancer therapy. In addition to human uses, endocrine disrupting steroids used in livestock operations contribute to widespread environmental contamination. Research by the US Geological Survey reported a high incidence of intersex fish in the Potomac watershed associated with sites of intense farming and high human population density; 75% of male smallmouth bass in the most densely populated heavily farmed Potomac basin had eggs in their testicles.

The issue of pharmaceuticals and personal care products in drinking water is not news to EPA, and yet despite the various safeguards and processes that EPA could have taken to develop a robust picture of the scope of the problem, the Agency has taken advantage of none of them. First, the Safe Drinking Water Act requires EPA every five years to publish a list of currently unregulated contaminants that should be considered for potential regulation. For these lists EPA has identified 130 potential chemicals for regulation – none of which are pharmaceuticals or personal care products. Second, in 1999 EPA promulgated an unregulated contaminant monitoring rule that imposed various monitoring requirements on community water systems for a list of unregulated contaminants; there are no pharmaceuticals or personal care products identified as unregulated contaminants for which water systems must monitor. Third, to our knowledge, EPA has never asked water systems to provide voluntarily testing and monitoring data for pharmaceuticals and personal care products. Fourth, the Safe Drinking Water Act requires community water systems to mail to each of their customers an annual report on the level of contaminants in the drinking water that they supply; there are no mandates to inform customers when pharmaceutical or personal care products are identified. Fifth, Congress mandated that EPA must address endocrine disrupting chemicals in drinking water; it has now been 12 years since this mandate and the Endocrine Disruptor Screening Program has failed to even start testing chemical contaminants.

The FY08 funding for the Toxics Program is $13.5 M; for FY09 the request is $10.7 M, reflecting a nearly $3 M cut from the USGS budget. The proposed cuts to the Toxics Program will significantly reduce research capacity on new and understudied environmental contaminants, including pharmaceuticals and personal care products.

Pharmaceuticals and Personal Care Products Are Contaminating Our Waterways and Sources of Drinking Water

An investigation by the Associated Press reported that pharmaceutical residues were detected in the drinking water of 24 major metropolitan areas across the country serving 41 million people.[1] Detected drugs included antibiotics, anti-convulsants, and mood stabilizer drugs. These results were supported by findings of the U.S. Geological Survey that sampled 139 streams in 30 states, and found organic wastewater contaminants and pharmaceuticals in 80% of sampled sites- including antibiotics, hypertensive and cholesterol-lowering drugs, antidepressants, analgesics,steroids, caffeine, and reproductive hormones.[2]

Pharmaceuticals and personal care products (PPCPs) include human and veterinary drugs, both prescription and over-the-counter, medical agents such as chemotherapeutic drugs and x-ray contrast media, antibiotics, anti-inflammatories, blood pressure and cholesterol lowering medications, psychotropic drugs, oral contraceptives, anti-seizure medications, fragrances, sunscreens, 'antibacterial' soaps, lotions, shampoos, and creams. They may end up in the environment through waste from human or animal excretion, improper disposal such as flushing down a toilet, runoff from animal feeding operations, or leaching from municipal landfills.

The problem of unintended movement of toxic and hormone disrupting compounds from pharmaceuticals and personal care products to wastewater effluents and drinking water sources is an international problem that has been documented and publicly reported by government experts and academic researchers for nearly two decades.[3] However, until recently, the public has been in the dark about the presence of these chemicals in our drinking water. As discussed more fully below, pharmaceuticals and personal care products have been excluded from the regulatory safeguards put in place by Congress – and the public's right to know has suffered as a result of that exclusion.

The contaminants come from many sources (medical waste, consumer waste, agriculture and industrial uses, etc.), have diverse toxicology profiles and biological activity, and are likely to have complex and poorly understood toxic interactions (antagonistic, synergistic, additive, etc.). However, among others, these contaminants share one very disturbing characteristic: in general, they are not effectively controlled under U.S. environmental statutes.

Widespread Antibiotic Contamination Poses a Serious Health Threat

Although the human health impacts of these exposures to pharmaceuticals and personal care products are poorly understood, what we do know is troubling. For example, we know that widespread exposure to antibiotics is contributing to the growth of bacterial resistance, and this problem is of grave concern. In the past several decades almost every bacteria that can cause infections in humans has developed resistance to at least one antibiotic, and some are resistant to multiple antibiotics. The Center for Disease Control has identified antibiotic resistance as one of the most pressing public health problems to face our nation.[4] Infections caused by bacteria with resistance to at least one antibiotic have been estimated to kill over 60,000 hospitalized patients each year.[5] However, antibiotic resistant bacteria are not limited to our healthcare settings. Methicillin-resistant *Staphylococcus aureus* (MRSA), a skin bacteria resistant to several antibiotics, was once found only in hospitals and nursing homes. MRSA is now commonly found in the community.[6]

Infection with MRSA can cause skin and soft tissue infections and pneumonia. Some MRSA infections are now only treatable with one antibiotic, vancomycin; it is extremely worrisome that microbial resistance to even this powerful antibiotic now has been reported.

Antibiotic resistance is caused by a number of factors including repeated and improper use of antibiotics in both humans and animals. Scientists also agree that exposure to low levels of antibiotics actually promotes bacterial resistance by exerting selective pressure for genes that promote resistance. Antibiotics end up in waste water because the body does not completely breakdown all drugs, so both the metabolized and unmetabolized drug are excreted by humans into wastewater. For example, when amoxicillin is ingested, 60-75% of the antibiotic is excreted unchanged into the urine. This antibiotic, now in the environment, may encounter other bacteria and promote resistance. It is unknown how much of an impact current low levels of antibiotics in drinking water are having on the problem of bacterial resistance. However, the potential has been recognized for many years.[7]

Massive quantities of antibiotics are used in agriculture both to treat infections and as food additives to promote growth and to compensate for conditions that contribute to infection. Animals raised in Concentrated Animal Feeding Operations (CAFOs) are at increased risk for infection due to close confinement and stress. In fact, it has been estimated that 70% of the antibiotics used in the US are for animal husbandry.[8] Improper and overuse of antibiotics in livestock and poultry can cause resistance in strains of bacteria that can infect humans. Furthermore, half of the antibiotics used in livestock are in the same classes of drugs that are used in humans. As a result the US Institute of Medicine (IOM) and the World Health Organization (WHO) both stated that the widespread use of antibiotics in agriculture is contributing to antibiotic resistance in humans.[9]

Large animal feeding operations generate a large amount of waste that can potentially contaminate groundwater and waterways contributing to antibiotic resistance and contamination of waterways with steroid hormones.[10] As occurs in humans, some portion of the antibiotics administered to livestock will pass unchanged through their bodies and will be excreted in their waste. It has been estimated that between 25-75% of antibiotics are excreted unchanged in feces and can persist in the soil after land application.[11] Manure is applied in large quantities as fertilizer in farm fields. In addition to potentially contaminating the food supply with antibiotic resistant bacteria, antibiotics in manure can persist in soil promoting the development of more antibiotic resistant bacteria. Animal waste and its associated contaminants can enter waterways through groundwater contamination, overflow of waste lagoons into surface water or by over-application of manure as fertilizer in farm fields. A recently published study found evidence of fecal contamination and increased levels of antibiotic resistant bacteria downstream of a swine concentrated feeding operation.[12] Other studies have found antibiotic resistance in groundwater underlying a swine waste lagoon.[13]

Widespread Contamination of Waterways with Steroid Hormones and Endocrine Disrupting Chemicals May Pose a Serious Health Threat

Congress and the EPA have recognized that some environmental contaminants are able to interfere with the actions of normal hormones.[14] These contaminants are called endocrine disrupting chemicals and have been demonstrated to interfere not just with sex hormones but other hormonal systems including the thyroid gland. Pharmaceuticals that mimic estrogen are excreted as waste by-products from the use of birth-control pills, menopause treatments, and cancer therapy. In addition to human uses, endocrine disrupting steroids used in livestock operations contribute to widespread environmental contamination.

Beef cattle raised in large feedlots are also treated with anabolic steroids to promote the growth of muscle. One of the most common steroids used is the androgen mimic, trebolone acetate. The breakdown products of this steroid are stable in water and bind with high affinity to the androgen receptor in fish. [15] Exposure to trebolone metabolites causes masculinization of female fish and cause reduced fertility at concentrations in the parts per trillion range. [16] Although we are much larger than fish, our bodies do not require larger doses of hormones to have effects. Sex hormones in all vertebrate species work in the parts per billion to parts per trillion range.

A recent study was done at an Ohio-based CAFO with a capacity for 9,800 cattle.[17] This study found detectable concentrations of trebolone in the discharge from the facility at levels that were sufficient to induce gene expression associated with exposure to androgens.

Recognition of the effects of endocrine disrupting chemicals in laboratory animal models has resulted in questions about whether problems in wildlife could be caused by environmental contaminants that are endocrine disrupting chemicals. For example, there is increasing public scrutiny and mounting evidence of the environmental effects of chemical contamination of waterways resulting in intersex fish in our nation's rivers and drinking water sources. Research by the US Geological Survey (USGS) reported a high incidence of intersex fish in the Potomac watershed at sites of intense farming and high human population density.[18] The USGS found 75% of male smallmouth bass in the most densely populated heavily farmed Potomac basin had eggs in their testicles. Other research has found environmental androgens associated with masculinization in female fish living downstream of pulp mills and concentrated animal feeding operations.[19] These organisms serve as sentinels for environmental contamination by endocrine disrupting chemicals and also raise concerns about potential impacts on human health, since humans share many of the same metabolic systems as fish.

There is concern that exposure to endocrine disrupting chemical contaminants could promote the growth of estrogen dependent cancers in some exposed people.[20] We must achieve greater understanding of which chemicals are causing these effects and conduct further laboratory testing to understand the potential human health effects.

Low Dose Exposures to Endocrine Disrupting Chemicals Pose Health Risks

The traditional toxicology dogma has been "the dose makes the poison" but when considering the toxicity from exposures to endocrine disrupting chemicals, the timing of exposure maybe more important than the dose. Exposures to endocrine disrupting chemicals during critical windows of development have been shown to have permanent effects.[21] Some of these effects, such as infertility or cancer, are not manifest until adulthood even though the exposure occurred during fetal or neonatal life. In animal studies, prenatal exposures to low levels (1 part per billion) of the synthetic estrogen, DES, has been shown to cause infertility, cancers of the reproductive tract and obesity.[22] Other laboratory studies have demonstrated that exposures to endocrine disrupting chemicals found in the environment, sometimes at low doses, are associated with reproductive harm or the development of reproductive tract cancers.[23] In humans, there is a noted decrease in the age at puberty, increases in infertility, increases in birth defects of male genitalia, increases in testicular cancer and continued high rates of breast cancer.

Each of these conditions has been shown in laboratory experiments to occur associated with exposures to endocrine disrupting chemicals during critical periods of development. We do not yet know whether similar exposures to endocrine disrupting chemicals in our environment could be causing the increased incidence of disease.

Despite the fact that low level contamination by pharmaceuticals and personal care products has been documented in the technical literature for some time, we have but an extremely rudimentary understanding of the hazards that these chemicals might pose to human health and the environment, or even the patterns of their occurrence in our nation's drinking water. We do know that many of these chemicals are designed to interact with our body's normal functioning, even at low doses, and that at therapeutic doses, drug interactions occur that can be harmful or even fatal. For example, we have no idea how low level exposure to psychotropic drugs might affect behavior of people and/or wildlife when occurring as a water contaminant. Therapeutic-level exposure to the anti-seizure medication valproic acid has been associated with increased birth defect risk,[24] but we do not know if exposures at lower levels will confer a similar risk. The long-term potential harm from the interactions of all these compounds, even at low exposures, is unknown, particularly for vulnerable populations such as infants, the elderly, and people with health ailments.

The Endocrine Disruptor Screening Program Has Failed to Even Start Testing After 12 Years

It is unknown how many of the pharmaceuticals and personal care products which contaminate our nation's drinking water are endocrine disrupting chemicals. We do know that some of the pharmaceuticals found in waterways are intentionally made to mimic our body's hormones. These include medications such as thyroid hormone replacement, estrogen replacement therapy, or birth control pills. Other pharmaceuticals are synthesized to block the action of our body's hormones, for example breast or prostate cancer therapies. We also know that certain chemicals found in personal care products are endocrine disrupting chemicals. For example, some chemicals used to carry fragrance, called phthalates, are known to inhibit the production of the male sex hormone, testosterone. However, for the vast majority of chemicals found as contaminants in the environment, there is no information to tell us whether or not they are endocrine disrupting chemicals.

In 1996, the Food Quality Protection Act (FQPA) required EPA to develop a screening program that would ascertain whether certain substances may have estrogenic (or other endocrine) effects on humans. The Safe Drinking Water Act extended this mandate to include substances that may be found in sources of drinking water. The Endocrine Disruptor Screening Program (EDSP) was intended to help define which chemicals could be capable of causing these effects and ultimately, provide information to be used to protect the public's health.

The Food Quality Protection Act mandated that EPA develop a screening program by 1998 and implement it by 1999. According to the original EPA timeline for implementation, the priority setting database was to be complete by June 2000 and was to be used for priority setting in November 2000. Despite the established deadlines, EPA failed to adhere to the original timeline for priority setting, which was the subject of prior litigation between NRDC and EPA. The Agency settled that lawsuit by agreeing to publish a proposed initial list of chemicals for screening by December 31, 2002 and to validate and begin requiring chemicals to be tested by December 2003. EPA missed these deadlines as well. In fact, it has been twelve years since Congress mandated EPA to create this screening program, but the Agency has yet to begin actual testing of any chemicals under this program.

EPA has failed to protect the public's health not only by delaying implementation of the testing program but also by refusing to regulate any chemicals that have been shown to be endocrine disruptors by other research studies. In addition to failing to reduce exposures to known endocrine disrupting chemicals, EPA has failed to begin testing the tens of thousands of other chemicals where we have no information about their potential endocrine disrupting effects.

A decade ago, the Endocrine Disruptor Screening and Testing Advisory Committee (EDSTAC) devised clear recommendations for a prioritization process for selecting chemicals for screening. These recommendations were adopted by EPA in 1998. For the first round of screening, EPA has proposed to test 73 chemicals that are in pesticides or are considered High Production Volume (HPV) pesticide inerts. EPA is under a mandate from Congress to begin testing by August 2008.

Although EPA finally appears to be making progress on this program, we are concerned that EPA has diverged from the EDSTAC's recommendations about how to select the chemicals for this initial list. In particular, and in light of concerns regarding pharmaceuticals and personal care products and other endocrine disrupting chemicals in water, EPA should include drinking water contaminants and not just pesticides in future testing protocols. Although EDSTAC recommended that EPA should test drinking water contaminants and mixtures of chemicals found in drinking water, EPA has not indicated that it intends to screen drinking water contaminants in the EDSP. In fact, EPA has the authority to do this under the Safe Drinking Water Act Amendments of 1996. These amendments authorized EPA to screen drinking water contaminants for endocrine disrupting effects. Specifically, Section 136 of the SDWA Amendments states, "In addition to the substances referred to in [the Food Quality Protection Act], the Administrator may provide for testing under the screening program authorized by [the FQPA] for any other substance that may be found in sources of drinking water if the Administrator determines that a substantial population may be exposed to such substance."

EPA's Failure to Regulate Pharmaceuticals and Personal Care Products as Drinking Water Contaminants

The issue of pharmaceuticals and personal care products in drinking water is not news to EPA. The Agency has been aware that these chemicals were contaminating our drinking water; studies on these chemicals ending up in our drinking water date back well over a decade. Yet despite this, EPA has failed to take the steps that would provide a fully, informed snapshot of the scope of the problem.

The Safe Drinking Water Act (SDWA) governs the regulation of contaminants in our drinking water supplies. Pursuant to the SDWA, EPA sets health-based standards for certain contaminants that may appear in drinking water. In addition to these regulated contaminants, the SDWA also created a system that would push EPA to determine whether other contaminants should also be regulated. Despite the various safeguards and processes that EPA could have taken to develop a robust picture of the scope of the problem of pharmaceuticals and personal care products in drinking water, the Agency has taken advantage of none of them.

First, the SDWA requires EPA every five years to publish a list of currently unregulated contaminants that should be considered for potential regulation. EPA is then required to make a final determination about whether or not to regulate at least five of the contaminants identified on the Candidate Contaminant List (CCL).

To date, the Candidate Contaminant List listing process has gone through 3 iterations, beginning in 1998 with the publication of CCL1 and then CCL2 in 2005. CCL1 contained 50 chemical contaminants, including industrial organic chemicals, pesticides, and inorganic chemicals; in July 2003, EPA decided not to regulate any of the nine chemicals it evaluated on the CCL1. CCL2 consisted of a subset of the chemical contaminants listed on CCL1; and in May 2007, EPA again decided not to regulate any of the 11 chemicals it considered from the CCL2. In February 2008, EPA published the draft CCL3. According to EPA, the 104 candidates for the draft CCL3 emerged from their evaluation of approximately 6,000 chemicals, 287 of which were pharmaceuticals.[25] EPA narrowed down the universe of 6,000 chemicals to a preliminary CCL3, which consisted of over 500 chemicals. In our review of the preliminary CCL3, aspirin was the only PPCP identified for further evaluation. According to the Associated Press, only one chemical on the draft CCL3 – nitroglycerine – is considered a pharmaceutical; however EPA indicated that it chose to include it on the draft CCL3 because of its use in explosives, not because of its pharmaceutical uses. All told, between the three CCLs, EPA has identified 130 potential chemicals for regulation – none of which are pharmaceuticals or personal care products.

Second, the Safe Drinking Water Act includes a process to inform both the determination about whether to regulate a contaminant on the Candidate Contaminant List and whether even to list a contaminant on the Candidate Contaminant List. Specifically, public water systems are tasked with collecting monitoring data on unregulated contaminants, and that data help EPA to decide whether or not to regulate a given contaminant. In 1999 EPA promulgated an unregulated contaminant monitoring rule (UCMR) that imposed various monitoring requirements on community water systems for a list of unregulated contaminants. The first round of the UCMR consisted of 26 unregulated contaminants that required some amount of monitoring data. The second UCMR identified an additional 24 unregulated contaminants not identified by the first UCMR. Just as with the three rounds of CCLs – there are no pharmaceuticals or personal care products identified as unregulated contaminants for which water systems must monitor.

Third, to our knowledge, EPA has never asked water systems to provide voluntarily testing and monitoring data for pharmaceuticals and personal care products.

Fourth, in recognition of the public's right to know the Safe Drinking Water Act requires community water systems to mail to each of their customers an annual report on the level of contaminants in the drinking water that they supply. These consumer confidence reports (CCRs) must contain, among other things, information on the source of the water, detections of regulated contaminants in the water, and levels of unregulated contaminants found in the water (those unregulated contaminants identified by the UCMR). Because none of these chemicals appear on any of the Candidate Contaminant List or Unregulated Contaminant Monitoring Rule lists, there are no mandates requiring water systems to inform their customers of the presence of these chemicals. As a result, the public's right to know about contaminants in their drinking water has been harmed by EPA's inaction.

Fifth, Congress mandated that EPA must address endocrine disrupting chemicals in drinking water. As discussed earlier, EPA has dragged its feet in implementing the endocrine disruptor screening program. Since so many pharmaceuticals and personal care products are potential endocrine disruptors, the Agency's failure to begin timely testing of chemicals continues to put all Americans at potential risk, simply from drinking water. Despite the recommendations of its advisory committee (EDSTAC) to include drinking water contaminants on its list of chemicals intended to be screened in the endocrine disruptor screening program as required by the SDWA, EPA's list of 73 chemicals does not include any pharmaceuticals or personal care products.

Sixth, EPA has yet to conduct a risk assessment of pharmaceuticals and personal care products in drinking water or assessed the health effects of the presence of these chemicals in our drinking water. EPA has failed in this respect, despite the fact that it claims it "established a leadership role [on the issue of pharmaceuticals and personal care products in drinking water] beginning in 1999 with publication of a critical review article that attempted to bring together the many different aspects of this complex issue."[26] Despite almost 20 years of evidence that these categories of chemicals were appearing in drinking water, EPA still has yet to conduct a risk assessment for pharmaceuticals and personal care products in drinking water.

Finally, as identified by the Associated Press report on pharmaceuticals in drinking water, not all water providers will test for these chemicals in their water. Those that do, rarely disclose this information to their customers. Until EPA chooses to include these chemicals on the UCMR – action EPA has so far refused to take – no disclosure is required under the SDWA and the public remains unaware of the presence of these chemicals in our drinking water.

USGS Budget Cuts Will Severely Impair Critical Water Monitoring Programs

We look to the federal government to protect the public's health. The public has reasonable and serious concerns that exposures to medications and other synthetic chemicals found in drinking water could be impacting human health. However, with limited data, it is impossible to gauge whether or not pharmaceuticals and personal care products in drinking water could be contributing to the development or progression of human health ailments. A commitment to rigorous long-term monitoring of our nation's waterways is absolutely essential for identifying contaminants, assessing risk, characterizing and localizing contamination patterns, identifying sources of contamination where possible, and measuring the effectiveness of mitigation measures. Resources to undertake this task are inadequate and falling, which will severely undermine our nation's ability to address this issue. Congress should allocate resources to undertake this task as a top priority.

The US Geological Survey (USGS) is responsible for the two main water-quality monitoring programs for the Nation's waterways: the National Water Quality Assessment Program (NAWQA) and the Toxic Substances Hydrology Program. Both water quality programs will suffer devastating blows if the proposed FY09 budget cuts are enacted. Adequate funding for these two programs is necessary to understanding many crucial aspects of water quality, including the impacts of new and under-studied contaminants such as pharmaceuticals and personal care products, and the efficacy of water quality policy-decisions. In short, the proposed deep cuts to these programs will strip the regulatory agencies of their ability to make wise decisions about how to allocate their limited resources to protect the Nation's waterways effectively.

The NAWQA is the larger of the two USGS water-quality monitoring programs; it monitors for environmental contaminants using established measurement methodologies for measuring (pesticides, volatile organic compounds, metals, etc.). Budget constraints over the last eight years has forced NAWQA to cut back from 496 surface-water fixed station water-quality monitoring sites in 2000, to only 113 sites in 2008. The FY09 proposed President's budget would reduce current funding by $9.8M, or 15% from the FY08 enacted funding. To adjust to this severe reduction, the USGS will be forced to cut monitoring at half of the existing ground water-quality sites so that it can continue to conduct research and analysis of existing data.[27]

The Toxic Substances Hydrology (aka Toxics Program) is the smaller of the two programs. It is a water quality research and methods development program that examines new and understudied environmental contaminants, such as the hormones, pharmaceuticals, and personal care products at issue in this hearing. The Toxics Program develops new capabilities, new methodologies, and new information that enable state water quality programs and NAWQA to address new issues effectively. FY08 funding for the Toxics Program is $13.5 M; for FY09 the request is $10.7 M, reflecting a nearly $3 M cut from the USGS budget. The proposed cuts to the Toxics Program will significantly reduce research capacity on new and understudied environmental contaminants, including pharmaceuticals and personal care products, as well as methyl mercury, arsenic, and nanomaterials.

EPA AND CONGRESS SHOULD ACT TO ADDRESS THE PROBLEM OF PHARMACEUTICALS AND PERSONAL CARE PRODUCTS CONTAMINATING THE PUBLIC'S DRINKING WATER

Under the Safe Drinking Water Act and the Food Quality Protection Act, EPA has the authority and obligation to ensure the safety of our drinking water. As discussed elsewhere in this testimony, EPA should:

- Include pharmaceuticals and personal care products in the unregulated contaminant monitoring rule to require public water systems to monitor for their presence in our drinking water and to identify in the consumer confidence reports the levels found in the drinking water; in the meantime, water systems should test for pharmaceuticals and personal care products and report the results to their customers;
- Add pharmaceuticals and personal care products to the candidate contaminant list 3 (CCL3) and evaluate the need to regulate the presence of these chemicals in drinking water;
- Immediately finalize and implement testing under the endocrine disruptor screening program and add drinking water contaminants, including mixtures of pharmaceuticals and personal care products to the list of chemicals that must be screened under that program;

- Evaluate and identify wastewater and drinking water treatment practices for removing pharmaceuticals and personal care products;
- In consultation with FDA and other federal research bodies, conduct studies to understand the health effects of discarded pharmaceuticals and personal care products on the nation's waterways and drinking water supplies; and
- Work with other federal agencies and states to prevent or limit the overuse of antibiotics in agriculture, particularly those that are critical for human use.

Congress needs to take additional steps to help address this issue, including:

- Establish take back programs for pharmaceuticals;
- Increase funding for wastewater and drinking water infrastructure; and
- Reform the Toxic Substances Control Act to reduce the number and amount of persistent, bioaccumulative and toxic chemicals that are released into the environment.

Thank you for inviting me to testify before you today. NRDC looks forward to working with the Subcommittee and Full Committee to address these important issues. I would be happy to answer any questions from the Committee.

End Notes

[1] Jeff Donn, Martha Mendoza, Justin Pritchard. Associated Press. March, 2008. http://www.msnbc.msn.com/id/23503485/

[2] Barber, L.B., Murphy, S.F., Verplanck, P.L., Sandstrom, M.W., Taylor, H.E., and Furlong, E.T., 2006, Chemical loading into surface water along a hydrological, biogeochemical, and land use gradient—A holistic watershed approach: Environmental Science and Technology, v. 40, no. 2, p. 475-486, doi: 10.1021/es051270q. (Supporting Information) http://toxics.usgs.gov/highlights/pharm_watershed/

[3] Aherne GW, Briggs R. The relevance of the presence of certain synthetic steroids in the aquatic environment. J Pharm Pharmacol 41:735-736 (1989).

Ankley GT, Brooks BW, Huggett DB, Sumpter JP. Repeating history: pharmaceuticals in the environment. Environ Sci Technol. 2007 Dec 15;41(24):8211-7

Kolpin, D. W.; Furlong, E. T.; Meyer, M. T.; Thurman, E. M.; Zaugg, S. D.; Barber, L. B.; Buxton, H. T. Pharmaceuticals, hormones, and other organic wastewater contaminants in

U.S. streams, 1999-2000: A national reconnaissance. Environ. Sci. Technol. 2002, 36, 1202-1211.

Ternes, T. A. Occurrence of drugs in German sewage treatment plants and rivers. Water Res. 1998, 32, 3245-3260.

Snyder, S. A.; Westerhoff, P.; Yoon, Y.; Sedlak, D. L. Pharmaceuticals, personal care products, and endocrine disruptors in water: Implications for water treatment. Environ. Eng. Sci. 2003, 20, 449-469.

Daughton, C. H.; Ternes, T. A. Special Report: Pharmaceuticals and personal care products in the environment: Agents of subtle change? (Vol. 107, p 907, 1999), Environ. Health Perspect. 2000, 108, 598-598.

[4] http://www.cdc.gov/drugresistance/healthcare/problem.htm

[5] Citation from Health Care Without Harm fact sheet, Antibiotic Resistance and Agricultural Overuse of Antibiotics.. Available at http://www.noharm.org/details.cfm?ID=938 &type=document

[6] Centers for Disease Control, Overview of CA-MRSA. http://www.cdc.gov/ncidod/dhqp /ar_mrsa_ca.html

[7] USA Today, in an 11/8/00 news article stated, "Experts fear that even low levels of antibiotics fouling the nations water supply may help create super-bugs: micro organisms that have evolved to survive an antibiotic's lethal assault."

[8] Mellon et al. *Hogging It: Estimates of Antimicrobial Abuse in Livestock.* Union of Concerned Scientists: Cambridge MA. 2000.

[9] Quotes from the Healthcare Without Harm factsheet. Antibiotic Resistance and Agricultural Overuse of Antibiotics. Novermber 8, 2005. Available at http://www.noharm.org/us/ food/issue

U.S. Institute of Medicine/National Academy of Science: "Clearly, a decrease in antimicrobial use in human medicine alone will have little effect on the current [antibiotic-resistant] situation. Substantial efforts must be made to decrease inappropriate overuse in animals and agriculture as well."

World Health Organization: "There is clear evidence of the human health consequences due to resistant organisms resulting from non-human usage of antimicrobials. These consequences include infections that would not have otherwise occurred, increased frequency of treatment failures (in some cases death) and increased severity of infections."

[10] Gilchrist MJ, Greko C, Wallinga DB, Beran GW, Riley DG, Thorne PS. The potential role of concentrated animal feeding operations in infectious disease epidemics and antibiotic resistance. Environ Health Perspect. 2007 Feb;115(2):313-6. Epub 2006 Nov 14.

Wallinga D, Mellon M, Roach S. Antibiotic use in swine farms in Alberta. Can Vet J. 2006 Dec;47(12):1153; author reply 1153-4.

Wallinga D. Public health advocate. Prev Vet Med. 2006 Feb 24;73(2-3):221-8. Epub 2005 Oct 28.

[11] Chee-Sanford, J.C., et al. Occurrence and Diversity of Tetracycline Resistance Genes in Lagoons and Groundwater Underlying Two Swine Production Facilities. *Applied and Environmental Microbiology*, April 2001. Vol. 6 no. 4, pp. 1494-1502.

[12] Sapkota AR, et al. Antibiotic-resistant enterococci and fecal indicators in surface water and groundwater impacted by a concentrated Swine feeding operation. Environ Health Perspect. 2007 Jul;115(7):1040-5.

[13] Chee-Sanford, J.C., et al. Occurrence and Diversity of Tetracycline Resistance Genes in Lagoons and Groundwater Underlying Two Swine Production Facilities. *Applied and Environmental Microbiology*, April 2001. Vol. 6 no. 4, pp. 1494-1502.

[14] http://www.epa.gov/endo/

[15] Durhan EJ, et al. Identification of metabolites of trenbolone acetate in androgenic runoff from a beef feedlot. Environ Health Perspect. 2006 Apr;114 Suppl 1:65-8.

[16] Durhan EJ, et al. Identification of metabolites of trenbolone acetate in androgenic runoff from a beef feedlot. Environ Health Perspect. 2006 Apr;114 Suppl 1:65-8.

[17] Durhan EJ, et al. Identification of metabolites of trenbolone acetate in androgenic runoff from a beef feedlot. Environ Health Perspect. 2006 Apr;114 Suppl 1:65-8.

[18] US Geological Survey (2008, February 11). Intersex Fish Linked To Population And Agriculture In Potomac River Watershed. *ScienceDaily*. Retrieved March 24, 2008, from http://www.sciencedaily.com /releases/2008/02/080208115302.htm

[19] Hotchkiss AK, et al. 2008 Fifteen years after "Wingspread" – Environmental Endocrine Disruptors and human and wildlife health: Where are we today and where we need to go. Toxicological Science Advance Access published Feb 16, 2008.

Durhan EJ, et al. Identification of metabolites of trenbolone acetate in androgenic runoff from a beef feedlot. Environ Health Perspect. 2006 Apr;114 Suppl 1:65-8.

[20] *State of the Evidence 2008*: The Connection Between Breast Cancer and the Environment Edited by Janet Gray, Ph.D., published by the Breast Cancer Fund and available at: http://www.breastcancerfund.org/site/pp.asp?c=kwKXLdPaE&b=206137

[21] Grandjean P, et al., The Faroes Statement: Human Health Effects Of Developmental Exposure To Chemicals In Our Environment. Basic Clin Pharmacol Toxicol. 2008 Feb;102(2):73-5.

[22] Newbold RR.Lessons learned from perinatal exposure to diethylstilbestrol. Toxicol Appl Pharmacol. 2004 Sep 1;199(2):142-50.

Newbold RR, et al. Perinatal exposure to environmental estrogens and the development of obesity. Mol Nutr Food Res. 2007 Jul;51(7):912-7.

[23] Vandenbergh JG. Animal models and studies of in utero endocrine disruptor effects. ILAR J. 2004;45(4):438-42.

Gray LE Jr, et al. Adverse effects of environmental antiandrogens and androgens on reproductive development in mammals. Int J Androl. 2006 Feb;29(1):96-108.

[24] Duncan S. Teratogenesis of sodium valproate. Curr Opin Neurol. 2007 Apr;20(2):175-80. Review.

James L, Barnes TR, Lelliott P, Taylor D, Paton C. Informing patients of the teratogenic potential of mood stabilizing drugs: a case note review of the practice of psychiatrists. J Psychopharmacol. 2007 Nov;21(8):815-9. Epub 2007 Sep 19.

[25] http://www.epa.gov/OGWDW/ccl/ccl3.html, last visited 11 April 2008; 73 Fed. Reg. 9627, 9652 (February 21, 2008).

[26] http://www.epa.gov/ppcp/faq.html#Whatwasepa, last visited April 11, 2008.

[27] USGS briefing sheet: Impacts of proposed FY09 budget cuts on National Water-Quality Assessment (NAWQA) program. February, 2008

In Pharmaceuticals in the U.S. Water Supply ISBN: 978-1-61942 -408-1
Editors: E. T. Blevins and S. E. Rush

Chapter 6

STATEMENT OF ALAN GOLDHAMMER, REGULATORY AFFAIRS PHARMACEUTICAL RESEARCH AND MANUFACTURERS OF AMERICA, AT THE SENATE ENVIRONMENT AND PUBLIC WORKS COMMITTEE SUBCOMMITTEE ON TRANSPORTATION SAFETY, INFRASTRUCTURE SECURITY, AND WATER QUALITY PHARMACEUTICALS IN THE NATION'S DRINKING WATER: ASSESSING POTENTIAL RISKS AND ACTIONS TO ADDRESS THE ISSUE*

April 15, 2008

Thank you Mr. Chairman and members of the Committee. My name is Alan Goldhammer, Ph.D., and I am the Deputy Vice President for Regulatory

* This is an edited, reformatted and augmented version of Statement given by Alan Goldhammer, Deputy Vice President, Regulatory Affairs Pharmaceutical Research and Manufacturers of America at the Senate Environment and Public Works Committee Subcommittee on Transportation Safety, Infrastructure Security, and Water Quality Pharmaceuticals in the Nation's Drinking Water: Assessing Potential Risks and Actions to Address the Issue, on April 15, 2008.

Affairs at the Pharmaceutical Research and Manufacturers of America (PhRMA), a trade association representing the leading research-based pharmaceutical and biotechnology companies. PhRMA member companies invested an estimated $44.5 billion in 2007 for innovative biomedical research to discover and develop new medicines that meet medical needs.

It is not a new development that drinking water may contain trace levels of pharmaceuticals as well as other household chemicals. What is new is that we now have more sensitive analytical methods to detect these substances at very low levels.

Reports of the presence of trace levels of pharmaceuticals in the water sparked the creation of a Pharmaceuticals in the Environment (PIE) Task Force within PhRMA in the late 1990s. The PIE Task Force is familiar with essentially all of the reports published worldwide of pharmaceuticals in water sources, including testing conducted by the U.S. Geological Survey (USGS), routine municipal water testing, testing by scientists from Europe to Asia, and even high school science students whose projects were on display in school science fairs.

PhRMA and its member companies advocate the safe and effective use of medicines and support the principles of product stewardship. As such, we are committed to applying the same level of scientific rigor to studying pharmaceuticals in the environment (PIE) as we apply to other areas of our business. PhRMA and its member companies have conducted research on a variety of PIE-related issues, including: 1) evaluating if detectable levels of pharmaceuticals in drinking water pose a risk to human health, 2) evaluating methods for the effective disposal of human medicines, and 3) determining the potential effects of human pharmaceuticals and their metabolites in surface waters on aquatic life. This research led to the publication of several research reports. We also continue to have an open dialogue with federal and state government officials and other stakeholders about the scientific findings. My testimony today will focus on the detection of trace amounts of pharmaceuticals, the potential human and environmental impacts, and how collectively we can minimize pharmaceuticals in the environment by proper disposal of unused medicines.

Pharmaceuticals are found in the environment primarily because trace amounts of medicines pass through the human body without being metabolized completely and make their way to surface waters through the municipal wastewater treatment system. These pharmaceuticals along with many other consumer products and household cleaning agents find their way into the wastewater from all households where there are family members who take medicines.

Where Do These Compounds Come From?

It is important for all of us to recognize that the concentrations of pharmaceuticals in the environment are extremely low. In fact, we probably would not be here today were it not for the development of improved analytical testing technology that has made it possible to detect trace amounts of consumer chemicals, including pharmaceuticals, in surface waters. The concentrations of pharmaceuticals reported in U.S. drinking waters are generally at trace levels of nanograms per liter (ng/l) or part-per-trillion (ppt). To put this in context, one ppt is about one second in 32,000 years or 1 penny in $10 billion. On average, those pharmaceuticals detected in U.S. drinking water are present at only 18 ppt.

It is not currently possible to prevent medicines from entering sewage. Wastewater treatment plants are designed to mimic the natural biodegradation processes that occur when organic compounds enter the environment. These systems are designed to reduce, but not eliminate, pollutants present in domestic wastewater. Therefore, a majority of the compounds used in households are expected to be present at trace levels in the discharges from wastewater treatment plants.

There are other minor sources for pharmaceuticals in the environment. Unused medicines can contribute to pharmaceuticals found in surface waters if they are flushed down toilets or poured down sinks. Recent Federal guidance on disposal recommends that medicines be disposed of in household trash or alternatively taken to local collection programs that accept unused medicine.

Is There an Effect on Human Health?

Many technical experts have contributed to the on-going scientific discussions about pharmaceuticals in the environment. The studies conducted to date suggest that it is highly unlikely that the very small quantities of even potent pharmaceuticals detected in the environment would be harmful to human health. Dietary exposure to hormones such as the estrogen that naturally occurs in milk and soy products is much higher than exposure to residues of any estrogen-like pharmaceutical in water. Trace levels of antibiotics found in surface waters are far below the concentrations necessary to develop antibiotic resistance in microbes. In summary, there appears to be no demonstrable risk to human health from detected concentrations of pharmaceuticals in surface waters.

Active pharmaceutical ingredients (APIs) are the most thoroughly studied substances in the world for their effects on human health. Billions of dollars are spent every year to discover and evaluate the efficacy and safety of new pharmaceuticals that can combat the effects of human diseases. Government agencies such as the U.S. Food and Drug Administration (FDA) and the European Medicines Agency (EMEA) have rigorous regulations that guide the collection and evaluation of information required to support the safety and efficacy of new pharmaceutical compounds. The FDA creates summaries of its evaluations of new APIs and makes them available to the public when the molecule is approved for use as a new medicine in the U.S. Acceptable use of an API in patients is judged by comparing the benefits from therapy with the risk of potential side effects. Approval of a new API routinely requires testing the efficacy and safety of the molecule in animal studies and directly in human clinical studies. Depending on the targeted use of an API, this testing normally includes evaluation of the molecule for properties of therapeutic efficacy, of acute, chronic and reproductive toxicity, and of genotoxicity and carcinogenicity. Additional evaluations of therapeutic efficacy and side effects are determined for patients in clinical settings. In order to understand systemic exposure from administration of an API, animal and human studies are conducted to measure the time course of absorption into the body, distribution into the blood and throughout the body, metabolism by organs and tissues, and excretion of the molecule and its metabolites via the liver and kidney into feces and urine. APIs and the metabolites eliminated from humans are piped to sewage treatment systems and then to surface waters. Detection of these chemicals has raised questions among some scientists about their safety to humans who might be exposed through drinking water.

Questions about the safety of drinking water have been addressed over several years by authors from the pharmaceutical industry, academia, government and the non-government sector. To date, no published investigation has found that exposure to these detectable residues creates a demonstrable risk to human health. A report for the Drinking Water Inspectorate of the United Kingdom concluded that even worst-case exposure for most major pharmaceuticals provided a high margin of safety.[1] Similarly, a study of human health risk assessments for 26 pharmaceuticals representing 14 general classes found in U.S. waters by the USGS was carried out.[2,3] The safety of drinking water and fish consumption was conducted using methods equivalent to those used by the U.S. Environmental Protection Agency (U.S. EPA) for drinking water under worst-case exposure conditions. The evaluation demonstrated that those detectable residues of pharmaceuticals in surface waters, and concentrations modeled under worst-case conditions, were safe and presented no demonstrable risk to human health.

An evaluation of the safety of drinking water in Germany concluded that risk is likely to be low from exposure to trace levels of pharmaceutical residues.[4] Another study evaluating the safety of potential residues of representative anti-cancer, lipid-regulating, anti-inflammatory, or analgesic drugs in drinking water found no demonstrable risk to human health.[5] A scientific paper published about a decade ago concluded that there was negligible human health risk from environmental exposure to an antibiotic that can be an allergen (phenoxymethylpenicillin), a potent estrogen agonist (17-alpha ethinylestradiol) or an anti-cancer drug (cyclophosphamide).[6]

For some perspective, researchers calculated that more than 90% of the 64 pharmaceuticals they evaluated would have a safety margin of at least 150,000 to reach a single therapeutic dose from water.[7] Put another way, over a 70-year lifetime, less than 20 percent of a single therapeutic dose would be ingested in drinking water. All of the APIs that they evaluated had a safety margin of 1000 or more to reach a single therapeutic dose from water. These safety margins should not be a surprise even for potent pharmaceuticals, since total use and excretion by patients are lower with higher potency pharmaceuticals.

Even though questions about the safety of detectable residues of pharmaceuticals in surface waters to human health have been evaluated in published articles for at least a decade, there are still understandable questions raised by the public about the possible impacts from the presence of these molecules. PhRMA and its member companies are committed to working with experts to openly consider and help answer these questions.

What Is the Potential Impact on Aquatic Life?

PhRMA believes that environmental impacts are already addressed through current regulations. The FDA requires an environmental assessment (EA) as part of the drug registration process in order to evaluate the potential for impacts to the environment (wildlife) as a result of patient use of a pharmaceutical. Substances that may enter the environment at less than 1 part per billion are typically excluded, although the FDA can require an EA for these substances based on extraordinary circumstances. Data on environmental fate (i.e., biodegradation, photolysis), transport (i.e., sorption, solubility), and potential effects (i.e., algae, invertebrates, fish) may be developed and submitted as part of the EA or as part of the drug registration process in other countries. The core finding of an EA is the relationship between the Predicted Exposure Concentration (PEC) and the Predicted No Effect Concentration (PNEC).

The presence of pharmaceuticals in surface waters has captured the attention of scientists, some of whom have speculated that the potential might exist for pharmaceuticals to impact wildlife. However, the scientific community and published literature believe that pharmaceuticals will not result in short term (acute) toxicity. This consensus is based on the documented low concentration of pharmaceuticals in the environment and a substantial quantity of acute toxicity data. PhRMA supports this view.

The results of acute and chronic toxicity testing demonstrate that a significant majority of pharmaceuticals do not present a significant risk from chronic exposure to the concentrations that exist in the environment (i.e., the PEC is well below the PNEC). Some compounds may, however, present extraordinary circumstances based on their potency or exposure scenarios. Compounds such as hormones are being investigated to determine whether extraordinary circumstances are warranted.

In a recent report the Associated Press (AP) claimed that "Drugs in Water Hurt Fish and Wildlife." This claim pointed to the potential for some of these situations to be considered extraordinary. Although the AP did not reference specific scientific reports that were the basis of these claims, most of the evidence cited by AP appears to be already known to scientists. In some cases, AP mentions preliminary results from ongoing studies. Without peer review or the ability to examine the methodology used and data produced from these studies, it is impossible to evaluate the science and validity of the AP claims.

The mere presence of a substance in water does not mean harm will result. The crucial factors determining whether a risk to wildlife exists are the concentration present in the water and to which an organism is exposed, as well as whether that concentration is at a high enough level to cause an effect. If the concentration is not high enough, that compound is likely to simply be part of the vast background of inorganic and organic chemicals, both natural and synthetic, present in natural environments such as soil and water.

WHAT IS PHRMA DOING?

When this issue first arose, PhRMA committed to understand the environmental significance of trace concentrations of pharmaceuticals in the environment and since that time has worked to assess the range of concentrations of pharmaceuticals that could be in the environment due to patients taking medicines and their potential effects on human health and aquatic life.

PhRMA developed the P*h*ATE® model to predict the concentrations of human health pharmaceuticals in wastewater treatment plant effluent, surface water, and drinking water throughout twelve watersheds in the U.S.[8] The model was recently upgraded to predict concentrations of APIs in biosolids that are removed from wastewater treatment systems. The P*h*ATE model is being used by researchers in Korea, Japan and more recently in Canada to predict environmental concentrations of APIs in the surface water of those countries.

PhRMA published an assessment of the human health effects of the APIs that were investigated by USGS during the national reconnaissance survey. In addition, PhRMA member companies have published dozens of articles in peer reviewed scientific journals evaluating the fate and effects of APIs in the environment.

PhRMA has developed the P*h*ACT® database to summarize all published English language peer reviewed literature about the effects of pharmaceuticals on aquatic life as well as treatment data for pharmaceuticals in wastewater and drinking water. Over 1,200 scientific papers have been entered into the P*h*ACT database through the end of 2007.

PhRMA representatives have participated in and led numerous scientific conferences over the past five years. PhRMA scientists also have chaired sessions on the PIE issue at several technical conferences, including a recent Water Environment Federation Symposium in Providence and the Society of Toxicology 47th Annual Meeting in Seattle.

With respect to unused medicines, PhRMA scholarship leads to the conclusion that the best action for the patient and the environment is for patients to take all of their medication as prescribed. Although the contribution of unused medicine disposal to the presence of pharmaceuticals in water is small, PhRMA encourages patients not to flush unused medicine and to dispose of them in an environmentally acceptable manner. In March 2008, PhRMA joined the American Pharmacists Association and the U.S. Fish &

Wildlife Service in launching the SMARxT DISPOSALTM program aimed at educating the public about not flushing or pouring any unused medicines down the drain. The SMARxT Disposal program is designed to raise public awareness and promote the use of household trash disposal or local collection programs as alternatives to drain disposal. Following these guidelines will also reduce the potential for abusing unused medicines where this may be a concern.

PhRMA has determined that both household trash disposal and incineration of unused medicines are environmentally acceptable ways to dispose of unused medicines. PhRMA is working with the University of Michigan's Graham Environmental Sustainability Institute to study the environmental impacts of take back for incineration and household trash disposal in order to better understand the overall environmental impacts of both these alternatives.

PhRMA is in the process of publishing a scientific study that finds if all unused medicines were placed in household trash and disposed of in municipal landfills, less than 0.1% of the total amount of medicine found in the environment would be contributed from landfills. In order to verify the conclusion of the landfill study, PhRMA is seeking the opportunity to partner with a researcher that is conducting landfill leachate sampling.

PhRMA is working with U.S. EPA, excelleRx, Inc. and other stakeholders to address the unique disposal issues associated with long-term care facilities (LTCFs). These facilities are required to witness destruction of unused medicines and which have typically relied on flushing down the toilet as the primary disposal method. PhRMA believes that approximately one-third of all unused medicines are generated by LTCFs.

Conclusion

PhRMA and its member companies remain committed to the ongoing study of trace amounts of prescription pharmaceuticals in the environment. Our PIE Task Force will continue to work with interested stakeholders to explore the scientific issues associated with pharmaceuticals in the environment. PhRMA will also partner with interested parties to better communicate the message about responsible disposal of unused medicines. As the available science demonstrates, and PhRMA concurs with, trace amounts of pharmaceuticals in the environment do not pose any human health or environmental risks.

REFERENCES

Watts, C., Maycock, D., Crane, M., Fawell, J., Goslan, E. (2007) Desk based review of current knowledge on pharmaceuticals in drinking water and estimation of potential levels, Final Report to Drinking Water Inspectorate of the United Kingdom, November 2007, Defra Project Code: CSA 7184/WT02046/DWI70/2/213.

Schwab, B.W.; Hayes, E.P.; Fiori, J.M.; Mastrocco, F.J.; Roden, N.M.; Cragin, D.; Meyerhoff, R.; D'Aco, V.J.; Anderson, P.D. (2005). Human pharmaceuticals in U.S. surface water: A human health risk assessment, Regulatory Toxicology and Pharmacology, 42, 296- 312.

Kolpin, D.W., Furlong, E.T., Meyer, M.T., Thurman, E.M., Zaugg, S.D., Barber, L.B., Buxton, H.T. (2002). Pharmaceuticals, hormones, and other organic wastewater contaminants in U.S. streams, 1999-2000: A national reconnaissance. Environ. Sci. Technol. 36, 1202-1211. [Including correspondence/rebuttal, 2002. Environ. Sci. Technol. 36, 4007-4008.]

Webb, S., Ternes, T., Gibert, M. and Olejniczak, K (2003). Indirect exposure to pharmaceuticals via drinking water. Toxicol. Lett. 142, 157-167.

Schulman, L.J., Sargent, E.V., Naumann, B.D., Faria, E.C., Dolan, D.G and Wargo, J.P. (2002). A human health risk assessment of pharmaceuticals in the aquatic environment. Hum. Ecol. Risk Assess. 8, 657-680.

Christensen, F.M. (1998). Pharmaceuticals in the environment – A human risk? Reg. Toxicol. & Pharmacol. 28, 212-221.

Webb, et.al. op. cit.

Anderson, P.D.; D'Aco, V.J.; Shanahan, P.; Chapra, S.C.; Buzby, M.E.; Cunningham, V.L.; DuPlessie, B.M.; Hayes, E.P.; Mastrocco, F.J.; Parke, N.J.; Rader, J.C.; Samuelian, J.H.; Schwab, B.W. (2004). Screening analysis of human pharmaceutical compounds in U.S. surface waters. Environ. Sci. Technol. 38, 838-849.

INDEX

E

F

G

H

I

K

L

M

N

S

T

U

V

W